Lonely Ideas

Lonely Ideas

Can Russia Compete?

Loren Graham

The MIT Press
Cambridge, Massachusetts
London, England

MIT Press books may be purchased at special quantity discounts for business or sales promotional use. For information, please email special_sales@mitpress.mit.edu or write to Special Sales Department, The MIT Press, 55 Hayward Street, Cambridge, MA 02142.

This book was set in Stone Sans and Stone Serif by the MIT Press. Printed and bound in the United States of America.

Library of Congress Cataloging-in-Publication Data

Graham, Loren
Lonely ideas : can Russia compete? / Loren Graham.
  pages cm
Includes bibliographical references and index.
ISBN 978-0-262-01979-8 (hardcover : alk. paper)
1. Technological innovations—Russia (Federation) 2. Technology and state—Russia (Federation) 3. Research, Industrial—Russia (Federation) 4. Industrialization —Russia (Federation) 5. High technology industries—Russia (Federation) I. Title.
T173.5.R9G73   2013
338.947—dc23
2013009425

10 9 8 7 6 5 4 3 2 1

To Pat and Meg
in gratitude for love and help

Contents

III   Can Russia Overcome Its Problem Today? Russia's Unique
Opportunity   143

Introduction

Russians, particularly in Soviet times, have often claimed that they invented many of the most important technologies of modern civilization: the steam engine, the lightbulb, the radio, the airplane, the transistor, the laser, the computer, and many other devices and machines. Western commentators have ridiculed these claims.

My recent research in Russian sources has revealed a big surprise. Russians did indeed build the first steam locomotive in continental Europe and the first operational diesel-powered locomotive in the world; they did first illuminate the avenues of major cities with electric lights; they did transmit radio waves before Guglielmo Marconi; they did build the first multiengine passenger planes (just a few years after the Wright brothers first flew); they did first create a new plant species through polyploidy speciation; they did pioneer in the development of transistors and diodes; they did publish the principles of lasers a generation before any others did; and they did build the first electronic computer in continental Europe. While the claims that they actually "invented" each of these devices are not correct, one thing remains clear: the Russians have absolutely legitimate claims to being pioneers in the development of all these technologies.

This recent research points to an important question, one impossible to ask without our new knowledge: if the Russians were pioneers in all these fields, why is Russia today so weak a player in world technology? The Russian economy is largely dependent on oil and gas; it is difficult to name a Russian high-technology manufacturer that is world class, with the possible exceptions of makers of weapons or space vehicles, and one or two software companies. Little Switzerland exports each year three or four times more high technology, measured in dollar value, than Russia. Hence the title of this book, *Lonely Ideas*. For three hundred years Russia has been adept in developing clever technical ideas and abysmal in benefiting from them. Lonely ideas, indeed.

Russians are a very creative people, as their achievements in music, literature, mathematics, and basic science richly demonstrate. In these fields Russians have deeply influenced the artistic and intellectual world. The educated public in America and other countries does not need to be told who Tchaikovsky, Tolstoy, or Dostoevsky were, and scientists and mathematicians similarly do not need reminding of the identities of Lobachevsky, Mendeleyev, Kolmogorov, or Landau.

However, Russians have been creative world leaders only in some areas of intellectual activity, not in all. Music, literature, mathematics, and some branches of basic science are largely creations of the mind, able to prosper as long as their practitioners are given good educations and the financial support necessary for them to work in their fields. Russians usually do well in such areas.

Technology, however, is a totally different subject; it is where the rubber meets the road, where intellectual creativity engages with society at large in a necessary and complex way, and where society can determine the success of technological projects, perhaps even unintentionally. The success of technology, which usually means profitability in a competitive international market, occurs outside the research laboratory, in the social and economic environment of the society as a whole. Russians have not done well in such activity. Where is the Russian Thomas Edison, Bill Gates, or Steve Jobs? Actually, they exist, but you have never heard of them because they fell flat on their faces when they tried to commercialize their inventions in Russia.

Walter Isaacson in his recent biography of Steve Jobs observes, "In the annals of innovation new ideas are only part of the equation. Execution is just as important."[1] Isaacson is correct, but his observation is incomplete, because he assumes a society where good execution by an individual can make the difference. In Russia even good execution was no guarantee of success because of the bad business environment.

A rich literature exists on the difference between innovation and invention, and that difference will help us understand the Russian problem with technology.[2] If we define "invention" as merely the development of new devices or processes, then the Russians are good inventors. If we define "innovation" to include the acceptance and implementation of new ideas, then we must conclude that Russians are miserable innovators.

The person who develops an idea with commercial potential needs a variety of sustaining societal factors if he or she is to be successful. These factors are attitudinal, economic, legal, organizational, and political. Society needs to value inventiveness and practicality; the economic system needs to provide investment opportunities; the legal system must protect intellectual property and reward inventors; and the political system must not fear technological innovations or successful businesspeople but promote them. Stifling bureaucracies and corruption need to be restrained. Many people in Western societies and, increasingly, Asian ones take these requirements for granted. Just how difficult sometimes they are to fulfill is illustrated by Russian history and by Russia today.

The curve of development of technology in many Western nations is a gradually ascending one with some flattening out, relative to other rising nations in Asia and elsewhere, in recent decades; that of China is like a giant "U" describing excellence centuries ago followed by a period of retardation during the era of Western imperialism, which was succeeded more recently by a sharp rise again. Russia's trajectory, on the other hand, for the past three hundred years has displayed an incredibly jagged line of peaks and valleys as excellence was followed by obsolescence, again and again. I refer to this spasmodic trajectory as Russian technology's pattern of "fits and starts." Russian technology currently is experiencing a hiatus in its fitful course and has fallen back on natural resources for economic strength.

When was the last time you went into an electronics or consumer technology store, picked up a gadget or consumer product that you want, turned it over, and noticed on the back "Made in Russia" (Sdelano v. Rossii)? Almost never. But how many times have you listened to a magnificent piece of music or read a great novel and found that its creator was Russian? Rather frequently. And this difference is not a recent one; it stretches back centuries. In that sense, Russian technology is very different from other areas of Russian art and thought. This book is an attempt to explain that curious fact. It is based not only on a study of the relevant sources but also on long-term residence in Russia, visits to dozens of Russian universities, research institutes, and industrial establishments, and conversations with thousands of Russian scientists and engineers.

I   The Problem: Why Can't Russia, after Three Centuries of Trying, Modernize?

1 The Early Arms Industry: Early Achievement, Later Slump

The Russian pattern of technological modernization in spurts followed by retardation was set very early, even in the seventeenth century. The rulers of Russia were eager to acquire powerful weapons that would permit them to wage war successfully against their neighbors, both defensively and offensively. As a result, they often directed the importation of foreign specialists, who created factories that were equal to the best standards of the day. The rulers hoped these arms facilities would maintain their excellence in later years, after the foreigners had returned home. The reason for disappointment was not, as foreign observers often thought, that the native Russians lacked technical ability—there were outstanding Russian arms craftsmen from the earliest days—but that social and economic retardation outside the factories gradually had their effect inside the shops, eroding the quality of the products. Such an effect was not unique to Russia, as is illustrated below by the examples of the Harpers Ferry and Springfield armories in the United States, where the effects of slavery and backward social development retarded production at Harpers Ferry compared to Springfield. In the Russian case, improvements at arms factories such as Tula could come only when the tsar ordered the next spurt of modernization, rather than through the authorities shifting, as the American government did, to favoring the progressive factories over the backward ones. This chapter concludes with a jump into the present, with a consideration of the AK-47, or Kalashnikov rifle, the most popular weapon of modern times.

The Tula factory has been improved to such an extent that not a single weapon factory in the world can compare with it.[1]

—An inspector of the Tula arms factory reporting to Tsar Nicholas I, 1826

I shuddered to see the devastation that the Minié carbines produced in the Russian columns . . . whose musket fire did not carry half the distance to the enemy as they pressed forward.[2]

—A British officer at the Battle of Inkerman, the Crimean War, 1855

The first quotation above would seem to indicate that in 1826, weapons production at the Tula arms factory, a leading armory of the Russian army, was the best in the world. The second quotation, twenty-nine years later, indicates that Russian troops in battle were using weapons that were entirely outclassed by those of their enemies at that time. How do we explain the difference?

The sad irony is that in 1826, the Tula arms factories *were* among the best in the world; however, in the decades immediately following, innovations made elsewhere in the manufacture of weapons were not implemented or matched by the Russian factories. Here is an example of the fitful trajectory that has so often characterized Russian technology. And here, as in so many other cases, there is a story behind this uneven development.

The Moscow Cannon Yard, established around 1479, used technology that astonished Western visitors. Originally tutored by Western foundry-men, the Muscovites developed their own procedures, which they kept secret from foreign visitors. Here were cast hundreds of heavy cannons for the Russian armies and some of the largest church bells ever made.[3]

In 1632, on orders of the tsar, the Dutchman Andrei Vinius established near Tula, south of Moscow, a forge for the manufacture of armaments that has had a continuous history to the present day. At first the Tula arms manufacturers employed the most modern methods. By the time of Peter the Great in the early eighteenth century, however, they were lagging behind Western European technology. Peter ordered the modernization of the factory, and especially the greater utilization of water power. He brought in Swedish, Dutch, Danish, and Prussian gunsmiths to teach Russian apprentices. Not only did he import foreign technicians, he also sent Russian mechanics abroad for education. One of them, Andrei Nartov, became a master machinist who developed lathes, mint presses, guns, and locks.[4] After Peter's death the policy of obtaining foreign assistance continued. Catherine the Great took an interest in the Tula factories in the last third of the eighteenth century,[5] actually helping to forge weapons during one visit. She also ordered that Russian gunsmiths be sent to England to improve their skills. During the Napoleonic Wars the Tula factories were major suppliers of guns of various calibers to the Russian armies.

The tsarist government in the early nineteenth century was very proud of its army and the weapons at its disposal. The army was the largest in Europe, one million men strong, and the Russian Empire had proved, after the defeat of Napoleon, that it was the dominant military power on the continent. In 1814 the Russians occupied Paris. Only British naval power counterbalanced the military forces of the tsar.

Realizing that this army had to be equipped with weapons equal to those of its potential adversaries, the tsarist government made a major effort to modernize its armory in Tula in the immediate post-Napoleonic period. In 1817 a master English gunsmith, John Jones, was brought with his family to Tula to manufacture gun locks by means of dies instead of the previous method of manual forging. He also introduced the use of drop hammers, and announced a program to produce interchangeable parts for his guns. By 1826 Jones had carried out such an impressive program of modernization that a tsarist inspector evaluated the Tula factories as being the best in the world.

Wishing to observe personally the wonderful progress in the Tula armory, Emperor Nicholas I decided to go to Tula on an official visit. He was told that at that time the armory had in its inventory 52,125 small arms produced by the new methods—a remarkable quantity. He was also told that no other country in the world was capable of producing such a large number of weapons with interchangeable parts. Nicholas actually visited the armory twice. On each occasion he randomly selected a few weapons from a number presented to him and asked that they be disassembled, the parts intermixed, and new weapons reassembled from the parts. The official report of his visits stated that the gun industry in the Tula factory had been brought, under the High Management of His Imperial Majesty, to "the highest degree of completion known at the present time."[6]

If this story is correct, it records a remarkable event. Historians of technology now agree that true interchangeability of parts for mass-produced small arms was not achieved elsewhere until the 1840s, when Americans in New England reached this goal.[7] Of course, many earlier claims in several different countries had been made, but under examination they failed to hold up to the criterion of true interchangeability. Where does Russia fit among these claims?

At this point the story becomes interesting, even paradoxical. We will probably never know exactly what happened during the tsar's visits to Tula in 1826, but evidence is mounting that he was deceived and that the guns manufactured in Tula at that time did not have truly interchangeable parts. The paradox, however, is that the evidence also indicates that the Tula factory in 1826 was indeed about as good as any large armory in the world. Some of its machines, such as the drop hammer and milling machines, were very impressive. The English master mechanic Jones, and perhaps some unknown Russian workmen, had made some improvements to the equipment Jones had used in England. Jones and his assistants achieved a greater perfection at Tula than Jones had known at home.

What, then, causes us to suspect that the tsar was deceived in 1826? And what evidence do we have that, despite this probable deceit, the Tula factories were roughly equal to the best armories in the world at that time, those in the United States and England? Historians of technology have examined surviving Russian small arms from the years 1812 to 1839 and have found evidence contradicting the Russian claims. The parts often bear the marks of hand filing, indicating that they were not interchangeable when originally produced but instead had to be laboriously and expensively fitted by hand. Some of the parts are even numbered, a practice not necessary if true interchangeability has been achieved. The U.S. historian of technology Edwin Battison examined this evidence in 1981 and noted that the gun parts were no more interchangeable than those of most guns produced in the United States in the same years. He observed, "Similar parts on similar guns made for the United States services at the same time would show comparable marked parts."[8]

Battison then went on to ask,

How is it possible that the Czar could have the incredible good fortune in his tests of locks and muskets? . . . Out of thousands of muskets it would be possible to select a small number that might be useful in such a demonstration. To select and prepare such exceptional muskets would be expensive, to place them so the Czar could innocently choose them, apparently at random, would be highly deceitful but it may have been done.[9]

Before we conclude that this sort of deceit was limited to Russia, we should notice that similar deceptions were evidently being practiced elsewhere at this time, most notably in the United States. In recent years U.S. historians of technology have destroyed the myth that Eli Whitney was the first developer of interchangeable parts.[10] In 1801, before a distinguished audience that included John Adams and Thomas Jefferson, Eli Whitney disassembled, intermixed, and reassembled the parts of ten gunlock mechanisms with no tool other than a screwdriver. Jefferson was so impressed that he wrote James Monroe that "Mr. Whitney has invented moulds and machines for making all the pieces of his locks so exactly equal, that take 100 locks to pieces and mingle their parts and the hundred locks may be put together as well by taking the first pieces which come to hand."[11] Jefferson's enthusiasm was understandable, for such guns could be easily repaired in the field.

We now know that Eli Whitney's claims were false. His muskets were not made of interchangeable parts. A historian who studied the physical evidence many years later concluded that the parts were "in some respects . . . not even approximately interchangeable."[12] Furthermore, Whitney was not able to achieve true interchangeability of parts during his lifetime, even though he remained a great exponent of the idea.

In 1826, one year after Whitney's death and the same year as the Tula demonstration, by remarkable coincidence three reports were published evaluating the state of small-arms manufacture in the United States and Russia. They allow us to compare the methods of manufacture in the two countries quite accurately. The three reports were the Carrington Report on the Harpers Ferry armory, an evaluation of Whitney's arms factory in Connecticut, and a report on the Tula armory in Russia.[13] The report on the Tula factory was the most detailed. The three reports, combined with physical evidence, show that guns with truly interchangeable parts were not being manufactured in large numbers in either country but that Russia was probably equal to the United States at that time in most operations and was superior in its ability to produce very large numbers of modern guns. Battison, who wrote the introduction to the Tula report, observed that "comparison of the scant few machines in Whitney's possession at his death with the number and variety in use at Tula . . . certainly punctures and deflates the overblown popular folklore enveloping Whitney."[14] John Hall at Harpers Ferry had produced 2,000 breech-loading flintlock rifles on an innovative interchangeable system that was promising for the future, but the Russians were already producing more than 20,000 guns a year that were comparable to the best U.S. muskets of that time.

And yet, starting from a position of approximate equality in the manufacture of small arms in the 1820s, in the course of the next thirty years Russia fell dramatically behind. From the 1830s through the 1850s, U.S. arms makers converted interchangeable-parts manufacture from an idea to a reality, and they expanded it into what became the American system of manufacture.[15] Russia missed out on this development. The gradual slippage of the Russian Empire was concealed for a while by the fact that the wars the Russians fought in the 1820s and the 1830s were against Turks and Caucasian mountaineers, whose arms were inferior. The revelation of Russia's military obsolescence came at mid-century, when on its home soil of the Crimea the Russian army tried to resist vastly better equipped British and French troops.

In the Crimean War the primary weapon of the Russian infantry was the smoothbore musket, many of which were produced at Tula. Some of the muskets were even flintlocks, since the program of conversion to percussion muskets initiated in 1845 was still incomplete. Many of these weapons were in poor repair, and their parts were not, by and large, replaceable. At the battles of Alma and Inkerman in the fall of 1854 the Russian troops faced French and British soldiers armed with rifles and minié balls, which had a lethal range approximately three times that of the Russian muskets. (Minié balls are a type of spin-stabilizing bullet originally developed by the French.) A Russian officer voiced his fear of the new weapons: "Seeing how in the Inkerman action, whole regiments melted from their rifles, losing a fourth of their men, . . . I am convinced that they will cut us down as soon as we fight in the open."[16]

How do we account for this slippage in quality of small-arms manufacture in only a few decades? There are several possibilities. As the historian of technology Edward Battison has suggested, perhaps the deception of claiming interchangeable manufacture in 1826 may have "inhibited further modernization in a country so autocratic that once the monarch had certified success . . . there was no way to pursue further appropriations for continued progress."[17] Another possibility is that Russia's diplomatic representatives abroad failed to report on progress being made in armaments manufacture in other countries.

Both of these answers may be partial explanations, but closer examination of the historical record yields another factor that seems to be the most important one. In some other leading countries the social and economic environment was one that nurtured and promoted technological development, while in Russia that same environment actually hindered such development. Without a social and economic milieu that independently promoted innovation, modernization of Russian technology could come only at those moments when the tsarist government suddenly noticed slippage and ordered reforms on the basis of a new importation of Western experts and machines. Such an abrupt rescue is what happened in 1817, when John Jones was brought to Tula, and it is what happened after the Crimean War, when the Russians again turned abroad for help with small arms. This time they paid attention to American-made rifles produced by the methods of interchangeable-parts manufacture. Russian military technology, like all technology in Russia and later the Soviet Union, advanced in fits and starts.

It is very easy to observe that social and cultural factors played important roles in the failure of the Tula armories to keep up with Western developments, but to make this hypothesis more persuasive requires a different level of proof. Do we have independent evidence of the influence of social factors on innovation in the field of weapons production in the first half of the nineteenth century?

A comparison of what was going on in the same years at two other national armories, those in the United States at Harpers Ferry, Virginia, and Springfield, Massachusetts, sheds light on these questions. Merritt Roe Smith's 1977 book on the Harpers Ferry armory is particularly helpful here.[18] Tula fell behind the leading Western nations in the production of high-quality small arms for at least some of the same reasons that Harpers Ferry fell behind Springfield in the same years.

In his careful comparison of the Harpers Ferry and Springfield armories, Smith concludes:

Harpers Ferry remained a chronic trouble spot in the government's arsenal program. Tradition-bound and often recalcitrant, the civilian managers and labor force accommodated themselves to industrial civilization most uneasily. The situation at the Springfield armory presented a marked contrast. There factory masters as well as mechanics seemed to embrace the new technology without any of the hesitancy or trepidation of their contemporaries in Virginia.[19]

Springfield, Smith continues, possessed an "expansive go-ahead vision," while the approach at Harpers Ferry was "myopic in quality—at once restricted in scope, static in attitude, and provincial in its processes."[20] As a result, Harpers Ferry fell behind in innovation, while Springfield and the Connecticut River Valley were the birthplace of the new system of production, an approach that spread from arms manufacture to the rest of U.S. industry.

Smith found the roots of the difference lay in attitudes toward modernization in the contrasting social environments of Harpers Ferry and

Springfield. Harpers Ferry was a small southern town in a predominantly rural area where a preindustrial ideology, craft traditions, and a social hierarchy influenced by slavery resisted the development of modern manufacturing methods. The town of Harpers Ferry was controlled by a few favored families, who were suspicious of social change. These privileged families could provide education for their children, but for other families, educational opportunity was extremely limited. A Lancastrian school existed in Harpers Ferry from 1822 to 1837, but then failed for lack of public support.

"The institution of slavery influenced social attitudes at the Harpers Ferry armory, even though few slaves were actually employed there. As Smith observes,

Because slaveholding heightened one's social standing in the community, the very possession of slaves by a few armorers made them extremely jealous of their rights and sensitive about the honor and dignity accorded their jobs. Any organizational or technical change that even slightly threatened to undermine their status was therefore firmly resisted. For instance, the most frequently repeated protests during the "clock strike" of 1842 accused the Ordnance Department of subverting the workers' freedoms and making them "slaves of machines." In a patriarchal society both terms conveyed deeply felt meaning.

Intensely proud of their special status as "master armorers," the veteran workers at Harpers Ferry resisted regimentation, machine work, time controls, and uniformity in production. In fact, when one superintendent, named Thomas Dunn, attempted in 1829 to enforce strict controls over work performance and quality of production, he was assassinated by one of the armorers. On other occasions the workers went on strike when they believed their status as craftsmen was being reduced to that of mere day laborers. "

The Springfield armory presented a sharply contrasting picture in the same years. There the majority of the workers had come from farms and villages in western Massachusetts, where they benefited from the public education system. As former free farmers who had never known slavery, they did not fear loss of their social status. Rather than oppose the adoption of machinery, they greeted its advent. Reared in a controlled environment in which the Puritan ethic still had force, they readily accepted the regulated environment of factory life. Springfield's workers were, compared to those of Harpers Ferry, remarkably disciplined and industrious. The managers of the Springfield factories were aggressive and devoted to technological progress.

Against the background of this contemporaneous example of the way in which the social environment influenced receptivity to technological innovation in the United States, let us examine the social context of the Tula armory.

By the first decades of the nineteenth century, the Tula arms factory was a complex of buildings and mills employing a great number of workers. It was, in fact, one of the largest armories in the world. In 1826 the Tula armory employed almost 14,000 workers, of whom more than 3,000 were members of a special estate (*soslovie*) recognized by the government and exempted from military service and from all taxes. Although technically serfs belonging to the state, these gunsmiths enjoyed remarkable privileges so long as they remained in their positions and obeyed the officers in charge of the armory. Some of them eventually came to own serfs of their own, a distinction particularly valued by men who were, legally speaking, serfs themselves.

The Tula works also employed 3,000 or 4,000 serfs belonging to land-owning nobility in the region. The gentry permitted these serfs to work off their estates as long as they paid a quitrent (*obrok*). Most were illiterate.

The history of the Tula ironworks is replete with conflicts between armorers and the rest of the peasantry in the Tula region. The armorers constantly struggled for special privileges that would distinguish them even more from the other peasants. Normally, all peasants were required to give their owners services in the form of labor or cash payments, and they were also required to pay taxes to the state and to serve in the army when called. The armorers of Tula gradually cast off most of these obligations. In their petitions to the tsar they reminded their ruler that they had special skills necessary for conducting wars, and they asked for freedom from obligations. These requests caused envy among the other peasants and sometimes open conflict. To avoid such confrontations, the tsarist government in the eighteenth century ordered the town of Tula to be split into different sections, with the armorers all living together in one settlement in which other citizens were forbidden to reside. The armorers did much of their work in their homes, and in peacetime they were allowed to produce tools, samovars, locks, and fittings, which they sold privately.

Despite the privileges they enjoyed, the Tula armorers were also tightly regulated. Without the permission of the government they could not leave Tula or abandon their profession. If they fled, they could be forcibly returned. In 1824 an armorer in Tula named Silin fled, was captured, was brought back to the armory, and was given two thousand strokes with a birch rod—surely an unsurvivable sentence. In the eighteenth and nineteenth centuries there were more than two thousand attempted flights from the Tula armory, averaging one a month.[21]

Although the Tula workers were economically exploited and suffered from poor living standards, they actually lived better than many other peasants in the empire.[22] But their lives were, as one Western historian described it, "hostile, violent, vengeful, quarrelsome, fearful, and vituperative."[23] The control of the owners over the serfs was bad enough, but even worse was the social oppression of serf by serf, which secured the system of authority. In Tula, when armorers became serf owners or serf supervisors, which many of the master armorers did, they usually soon gained reputations as extremely severe taskmasters.

The most prestigious workers in the Tula armory were the master armorers, who produced richly ornamented guns, for which Tula became famous (some Tula guns are valued collectors' items today). These artisans specialized in such techniques as sunken relief, damascening, inlaying, chasing, blueing, and the carving of hunting scenes on the stocks of the guns. Such armorers considered themselves artists, not workers, and some of them, such as Petr Goltiakov, became famous. Goltiakov started out as an ordinary gunsmith, but as a result of his outstanding talents he became lock overseer of the entire armory. In 1852, just before the outbreak of the Crimean War, he was appointed purveyor of guns to the Grand Princes Nikolai and Mikhail, a position of great honor.[24] Goltiakov's beautiful guns were "presentation arms" of the type made for the ruling family and highest officers, not arms made for the infantry. The system of rewards for armorers at Tula was heavily skewed in favor of craftsmen like Goltiakov producing presentation weapons, not machinists making ordinary arms. As a result, a few of the best Russian guns were very good, but the average were of poor quality.

The master craftsmen of Tula relied on their personal skills, and they resisted any innovation that would reduce their status to the ranks of state peasants, where most of their families had started. Tula had a long history of resistance to machines.[25] During the time of Peter the Great, the master armorers submitted a protest to the Russian Senate against the introduction of water power. For decades the armorers resisted the transfer of work from their homes, where many of them had small forges, to centralized plants. In 1815 the tsarist government installed a steam engine in one of the buildings in Tula, but an official report of 1826 states that the engine was still not being used. At that time most of the armorers worked at home at hand labor rather than in the main shops. Even as late as 1860 only 35 percent of the Tula armorers worked "within the walls of the factory" rather than at home.[26] For many of them, moving to the factory would mean becoming industrial employees instead of artists.

In 1851 the Crystal Palace Exhibition was held in London, a grand affair where many nations displayed products of art and industry. The Russians, along with Austrians and the French, presented artistic displays in which the artifacts were heavily ornamented. The U.S. display was strikingly different. As the historian Nathan Rosenberg observed, "A visitor who came to the American exhibit to gratify his aesthetic sensibilities was wasting his time. It was severe and utilitarian in nature."[27] Critics soon called the U.S. exhibition area the "prairie ground." Among the items on display were ice-making machines, corn husk mattresses, railroad switches, and telegraph instruments. More to the point, the display included small arms produced by the firms of Samuel Colt of Connecticut and Robbins and Lawrence of Vermont. The Robbins and Lawrence weapons were made of truly interchangeable parts. Colt announced that he relied on machine production because "with hand labor it was not possible to obtain that amount of uniformity, or accuracy in the several parts, which is so desirable."[28] This 1851 display was a warning for the Tula armorers. The future, so bloodily displayed at the battles of Alma and Inkerman in the Crimea just three years later, was one in which art was to be lost, but lethal uniformity was to be gained. The Russian army was humiliated at Crimea by the superiority of the British and French rifles, even though only a few decades earlier Russian weapons had been equal to the best of their adversaries' arms.

The story of early achievement and subsequent decline that we see at the Tula arms works is a common cyclical pattern in the history of Russian technology. I discuss different episodes of this cycle in different industries in the following chapters of this book. The pattern was repeated under the tsars, under Stalin, and under Brezhnev, and it is recurring again in post-Soviet Russia. At present, the one area of economic development in which Russia continues to be a leader is oil and gas production, but even here, Russia has not kept up with the new technologies for extracting oil and gas that have emerged in other countries (although, once again, the Russians developed fracking technology early on but failed to follow up).[29]

Can Russia ever escape this deadly cycle? In principle, there is no reason why it cannot, but in practice the escape has proved to be extremely difficult. The example of Tula and why it fell behind is still relevant to Russia today.

Postscript: The AK-47

Although it is a jump into the twentieth century, in this discussion of small-arms production in Russia it is appropriate to add something on the Kalashnikov rifle, or AK-47, the most popular, widespread, and numerous hand weapon of the last sixty years.[30] At first glance the Kalashnikov seems to be an exception to the story told here about the difficulties of Russian technology. After all, it is the most famous, reliable, sturdy, simple, and inexpensive automatic rifle in history. It is the weapon of choice of many armies, guerrilla groups, rebels, and terrorists. For its purposes, it is a brilliant technology. Along with *Sputnik*, it is probably the best-known Soviet technological product. Millions of people instantly recognize it because of its distinctive curved magazine, raised sight, and gas-chamber tube.

Rather than being an exception to the thesis of this book, the AK-47's history is a further illustration, for the following reason. Russia has received precious little economic benefit from originating the most widely manufactured firearm in the world. The company in Russia that makes it, the Izhevsk arms works, has recently been threatened with bankruptcy. Most AK-47s in the world, however, are not manufactured in Russia but are knockoffs produced without any payment to Russia. In February 2012 the vice-premier of Russia, Dmitrii Rogozin, complained that such unauthorized AK-47 producers "pay us nothing. . . . This undermines our export position."[31] By late 2012 approximately 80 percent of the few Russian exports of AK-47s were to the United States, where it has become a cult weapon for collectors who want the original Russian product rather than an imitation.[32]

The man usually credited with inventing the rifle, Mikhail Kalashnikov, did not make any money from its manufacture (although he did receive a Stalin Prize and great recognition).[33] The Izhevsk arms works did not take out a proper patent on the rifle until 1999, more than half a century after its development (the term AK-47 refers to the date of its design, 1947, and

"AK" is an initialism for Avtomat Kalashnikova). By that time the design of the AK-47 was in the public domain. Without the financial bonanza that a proper control over rights to this famous gun would have brought its manufacturer, the Izhevsk arms works has not been able to keep up with further advances in small-arms production and is losing its standing. The AK-47 remains very popular for close-in combat, but it has a short range (350 meters) and lacks the accuracy of modern automatic weapons. (There are plans for its modernization, sixty-five years after its first design.) All this fits in with the thesis of this book: throughout the centuries Russia has produced some remarkably good technologies but almost never derives appropriate economic benefit from them and usually fails to keep up with later developments in the same technological area. As time goes on, the AK-47 increasingly fits the pattern of fits and starts in Russian technology, although the start in this case was exceptional in its brilliance and international recognition.

2   Railroads: Promise and Distortion

Russia was an early leader in the development of railroads. In 1835 a Russian father-and-son team, the Cherepanovs, produced a steam locomotive that could pull a sixty-ton load. The Cherepanov locomotive was not developed further, however, and Russia soon had to turn to foreign powers, the United States and other European nations, for the construction of railways. Nonetheless, Russia was one of the handful of pioneering nations in the construction of railroads. The first Russian steam railway was opened to the public in 1837, the same year as the first one in Austria and only five years after the first steam railway in France. St. Petersburg and Moscow were connected by rail before Chicago and New York.

Russia's prominence in railroads did not last long, however. The railway network in Russia expanded much more slowly than in Britain, France, Germany, or the United States. Between 1844 and 1855 no new railroads were built in Russia.[1] These were years in which railroad construction was booming in other industrializing countries. By 1855 Russia had only 653 miles of railway, compared to 17,398 in the United States and 8,054 in England.[2] Why was Russia progressing so slowly in railroads, despite its early promising start? The explanation is largely political and economic.

Production is about two engines a month ahead of any establishment in any country that we have knowledge of.[3]

—George Washington Whistler, American railroad engineer, speaking in 1847 of the St. Petersburg railroad factory

In 1833 a Russian craftsman named Miron Cherepanov came to Newcastle, England, to visit the railway factory of George Stephenson.[4] Both Cherepanov and Stephenson were pioneers in developing steam engines and locomotives. Stephenson (with his son Robert) and Cherepanov (with his father

Efim) would build dozens of steam engines in their lives. Just four years before Cherepanov visited the Newcastle factory Stephenson had demonstrated the *Rocket*, often called the first successful train engine in history.[5] Back home in Russia, Cherepanov would in the following year, 1834, build the first locomotive on the European continent to be made in its own country and not imported from England. It was successfully demonstrated and had a 30 horsepower engine. A second model followed a year later with a 46 horsepower engine. (Stephenson's *Rocket* was 20 horsepower.)

There were some similarities between Cherepanov and Stephenson. Both came from poor mining families. Neither Cherepanov nor Stephenson had an advanced education, and both were weak in mathematics and in the thermodynamic theory that lay behind their locomotives. Both were highly skilled metal craftsmen and built engines that were advanced for their time. Yet there was a big difference in their fates as entrepreneurs. Stephenson's inventions were widely duplicated and had an impact on the entire later development of railways. Stephenson is part of the standard history of technology and is often referred to as "the father of railways." Cherepanov's inventions were not duplicated, and today he is largely forgotten. Why this difference?

Although Stephenson came from a very poor family, he took out patents on his locomotives and managed, with the help of investors, to establish his own firm, Robert Stephenson and Company (named after his son). His company sold its services and products to a wide variety of customers, who used the engines for many purposes: hauling ore from mines to smelters, pumping water out of mines, transporting passengers on the newly constructed Liverpool and Manchester Railroad. In short, Stephenson's engines became commercial products.

Cherepanov was a serf, bonded to his masters, the Demidov family, which owned mines and smelters in the Ural Mountains. Cherepanov's entire life was spent in their service, and he was totally dependent on them. A patent system did not exist in Russia, and even if it had, Cherepanov, as a serf, would not have been eligible to apply for one. His supervisors sent him to England not because they wanted him to find out about locomotives; they dispatched him there to learn why the market for Russian iron was decreasing and what the secret was behind the rapidly improving quality of English iron. The directors of the Demidov factories in the Urals were not interested in locomotives; they thought it was cheaper to haul the ore from their mines to their smelters using large numbers of cheap bonded serfs for labor, assisted occasionally by horses. The incentive and the vision for employing locomotives were missing in Russia.

Stephenson's locomotives in England were used for commercial purposes in mines, for transporting cotton to textile mills, and, on the first public railways, for moving passengers and freight. Their owners wanted to make money. Russia was probably the first country outside England with a working steam engine, but it was used for pumping water for the exuberant fountains of the tsarina's summer palace in Peterhof. The first railway in Russia, built in 1835–1837, ran between the ruler's winter residence in St. Petersburg and a station near this same ruler's summer residence—the Tsarskoe Selo Railway. Both the water-pumping steam engine in Peterhof and the railway between the royal residences were built largely for show and the rulers' pleasure, not for commercial advantage of the sort Stephenson's company was seeking in England. The extravagance and wealth of the Russian tsars were fully demonstrated, making the stingy economic calculations of the English railway builders seem dismal by comparison. Yet the shrewd English industrialists were augmenting the power of the British Empire, in contrast to the demonstrative displays of the Russian rulers, which did nothing to increase their actual power. When the Tsarskoe Selo Railway was built, the engines were purchased from Western Europe, even though Cherepanov had already produced domestic Russian locomotives. He had not been able to improve them through ongoing tinkering and changes, the way Stephenson and other English engineers did, because there was no demand from the Russian market.

In this contrast between Stephenson's and Cherepanov's fates as engineers and producers, it is evident that the most important factors in promoting technological progress are not the invention of the device itself (neither Stephenson nor Cherepanov was the "inventor" of steam engines or of locomotives; both had predecessors) but the social and economic stimuli that cause a certain technological innovation to be picked up and propelled further.

Eventually, in 1842, Tsar Nicholas I, impressed by the first railroads he saw in Western Europe, ordered that a railway be built between Russia's two major cities, St. Petersburg and Moscow. Many of his advisers and ministers opposed the decision, but there were exceptions. One was Pavel Mel'nikov, a brilliant railroad engineer, properly educated in engineering science and experienced in "railroad studies" in Europe and America.[6] Russia had another chance to be a leader in the construction of railroads. Mel'nikov favored hiring one of his American acquaintances, George Washington Whistler, as a consultant, especially on rolling stock. Whistler, working with Mel'nikov, engaged the American firm of Winans and Harrison to construct railway engines and cars in St. Petersburg, and soon the factory there became a model of its type. Harrison bragged that the production of his St. Petersburg railroad factory "is about two engines a month ahead of any establishment in any country that we have any knowledge of." For a brief moment, the tsarist empire possessed one of the best-equipped railways in the world.

Mel'nikov pleaded with the tsar to construct a wide network of state-controlled railroads, emphasizing their economic benefit. But once the St. Petersburg–Moscow railroad was built, Russia once again fell behind while railroads spread rapidly in Europe and America. It is true that the tsar continued to favor railroads in principle, and commissioned a number of studies of further construction, but very little was actually done. Building railroads was extremely expensive, and Tsar Nicholas did not favor allowing private investors to finance them; he, like Mel'nikov, wanted state control. By 1855 Russia had only 653 miles of railway, compared to 17,398 in the United States and 8,054 in England.

As was the case with armaments at Tula, the wake-up call came with the Crimean War in 1854–1855, when Russia's army, fighting the British and French on the Black Sea, could be supplied only by horse-drawn wagons pulled over muddy roads that were nearly impassable in the spring and fall. No railroad existed south of Moscow. Mel'nikov saw his chance, and wrote a stirring memorandum for the new, reforming tsar, Alexander II.[7]

He started by pointing out to the tsar that only twenty-five years had passed since the opening of the first railway from Manchester to Liverpool but that already the larger parts of Europe and the United States were now served by railways. This strikingly rapid development showed, he observed, that these Western countries were aware that railways were absolutely necessary if a nation did not wish to fall behind in its economic development. In taking advantage of railways he saw the United States as the leader, noting what he called the "colossal" way in which the federal government and private capital worked together to advance the new network of transport. And he noted that the United States and Russia were similar in the vastness of their territories, their climates, and their need for transportation to newly settled regions. But, he continued, Russia had some advantages over the United States: its topography was flatter and less mountainous, and with its centralized government it had the potential to create a more rational railway system. He criticized the United States for the different railroad policies of the individual states and the inefficient competition of the private railway companies.

Mel'nikov maintained that Russia had an even greater need for railways than the United States since, in addition to the similar desire to develop natural resources, agriculture, and industry, Russia had to defend itself from many potential enemies on land and had to be able to dispatch troops wherever needed. Here he was obviously referring to the debacle of the Crimean War. Imagine, he urged the tsar, a situation in which Russia had a network of railways permitting it to throw troops in almost any direction—to the Baltic Sea, to the Black Sea, to the Caspian Sea—while maintaining central control and central barracks in Moscow and St. Petersburg. The army would then be so much more effective than in its present situation that it might be possible to make do with a smaller army (at the time, Russia had the largest army in Europe, maintained at great expense). He urged the Russian government to get busy with the construction of railways "without delay." Every day of hesitation, he warned, would come at a high cost.

Railways, Mel'nikov continued, were being built elsewhere in a variety of different ways. In Great Britain he saw the major role being played by shareholding private companies without much government involvement; in Belgium and parts of Germany, he thought that the governments were building railways largely alone; in France and in parts of Germany he saw the governments participating together with private companies in the financing of railways. In the United States he saw the dominant role being played by private companies, but noted they were helped enormously by positive government policies, especially through land grants.

Mel'nikov believed the superior model was a combination of government and private initiatives, with the government retaining ultimate control. The upper hand of the government was particularly necessary in Russia, he maintained, for political and military reasons, and because of the relative lack of private capital. And he thought that quality and rationality were best regulated by the government, relying on qualified specialists in governmental institutions (such as Mel'nikov). He then sketched out a system of railways for Russia that would connect the center with areas on its northern, western, and southern borders, as well as with its mines and industries in the Urals.

Tsar Alexander agreed, and so once again a spasm of modernization, expressed in railroad building, gripped Russia. The tsar was now willing to let private investors, including foreign ones, help finance the railways. However, the expansion of railways that occurred was still much slower than in Europe or America. The tsarist government retained tight control.

A new leap forward, once again under state control, would come decades later under the direction of a new bureaucrat, Sergei Witte, and with a new tsar's support. Witte was the driving force behind the Trans-Siberian Railway, which he saw as the key to a new economic system.[8] A follower of the economic views of Friedrich List, Witte emphasized industrial growth, control over tariffs to the benefit of the state, the expansion of technical education, and promotion of growth through improved transportation, especially railways.

Yuri V. Lomonosov was a talented railroad engineer who benefited from Witte's expansion of engineering education and became a pioneer in railroad locomotive construction.[9] Lomonosov was a controversial character with a very interesting life: a member of a noble family, he was also a political revolutionary and played radical roles in both the Revolution of 1905 and the February Revolution of 1917. After the later Bolshevik Revolution in October and November 1917, the leader of the new Soviet state, Vladimir Lenin, considered Lomonosov for the high position of minister of railways and met with him several times. Lomonosov, an inordinately ambitious man, would have loved the position but eventually lost out because of his "inappropriate" background.

Lomonosov continued to work for the Soviet state and in 1924 created the first operational mainline diesel-powered locomotive in the world, which entered service in 1925. Many people in other countries had attempted to create such a locomotive, since diesel engines are much more thermally efficient than steam engines, but Lomonosov was the first to develop one that was a practical success. This achievement won him fame in Western countries, even more than in Soviet Russia itself. In the young Soviet state, however, he was looked upon by the increasingly militant Soviet regime, in which Stalin was gaining greater influence, as a "bourgeois specialist," a man who could not be trusted because of his social background. Furthermore, he was an outspoken man who easily made enemies.[10]

In retrospect it is clear that, however good Lomonsov may have been at the design, testing, and operation of locomotives, he was a hopelessly confused man, even naïve politically. On the one hand he considered himself a Marxist and a revolutionary; on the other hand he gladly accepted high positions in the administration of tsarist railways, becoming so prominent he was even a possible candidate for assassination by his own revolutionary comrades as a tsarist state official. At the same time that he was administering railways for the tsarist government he was helping revolutionaries in 1905 to build bombs. After the 1917 Bolshevik Revolution he called himself a "non-Party Soviet Marxist," refusing to join the Communist Party, a requirement for acceptance into the higher ranks of Soviet administrators. As a result, many people in both tsarist Russia and the Soviet Union did not trust him, and the distrust extended across the political spectrum. To many of his revolutionary comrades before the Revolution he was a part of the tsarist establishment. To Soviet authorities after the Revolution he was a bourgeois specialist capable of betraying the revolutionary cause, despite his constant protestations that he was a loyal Marxist supporting the regime. His refusal to join the Communist Party made his statements of loyalty seem insincere.

Criticism of him mounted from radical circles, and he gradually realized that he was headed for trouble in the Soviet Union. Therefore, in 1927, just before the Soviet regime began arresting hundreds of such bourgeois specialists, Lomonosov seized an opportunity to emigrate by remaining in Britain, where he was temporarily working on a mission supported by the Soviet government. He became a British citizen and later briefly a professor in the United States.

When he emigrated to Britain and the United States to avoid arrest in the Soviet Union, he alienated many people who could not understand his hostility, as a loyal socialist, to private enterprise. He was a singular, hulking, strong-tempered, fiercely independent man. Even in his own family the nickname given him, hopefully affectionately, was "the monster." Yet his talent as a developer of locomotives was unquestionable. (All these characteristics of Lomonosov are beautifully described in Anthony Heywood's recent biography, *Engineer of Revolutionary Russia: Iuri V. Lomonosov (1876–1952) and the Railways*.) If he had had an opportunity to start his own company—something he would never do because of his scorn for capitalism—or to work for a government or a private company that valued his talents and overlooked his eccentricities, he might have been a great success. But that was not to be.

The distrust that Lomonosov engendered was one reason his diesel locomotive did not receive the further refinement and development in Russia that it deserved, although Soviet railways eventually used many diesel engines. Working on Soviet contracts, Lomonosov developed his first diesel locomotive in shops in Germany because he could not find suitable facilities in the early twenties in Soviet Russia, a country undergoing economic and political crisis. His reasons for working in Germany instead of Russia were technically rational but politically dumb, and his long presence there deepened the distrust of his Soviet critics, who suspected, incorrectly, that he favored German capitalists over Soviet communists.

As a result, instead of benefiting from Lomonosov's innovation, the Soviet government purchased, at great expense, locomotives from Sweden and Germany. Soviet Russia in the 1920s spent about 30 percent of its gold reserves on thousands of foreign-made locomotives and freight cars.[11] In later years Soviet engineers working in their own factories often copied foreign models of locomotives. For example, several of the most popular Soviet locomotives, the TEM2, TEM3, and TEP70, were based on copies of locomotives made in the United States by the American Locomotive Company, or ALCO, and on the 1967 British Rail HS4000, a prototype mainline diesel locomotive named the *Kestrel*. And the pattern of the influence of foreign technology on Russian railways has continued down to the present. The impressive fast electric train, the Sapsan, that today runs from Moscow to St. Petersburg was built by the German company Siemens. In November 2012 Russian Railways signed an agreement with Siemens to buy 675 locomotives at the cost of several billion dollars.[12]

Russia, belatedly and by way of periodic spasms of expansion, came to possess an impressive system of railroads, but its railroad-building trajectory was shaped less by economic forces than by the central power of the government, whose enthusiasm for railroads and railroadmen waxed and waned and which sometimes built railroads in ways and in places they never should have been built. The capstone of this "building by edict" came with the construction of the Baikal-Amur Mainline in the late Soviet period, a gargantuan construction project and an episode of striking irrationality. That strange story is considered in chapter 5, on Soviet industrialization.

After the collapse of the Soviet Union, post-Soviet Russia tried once again to modernize its railways in joint ventures with foreign companies, such as Siemens and the Canadian firm Bombardier. The fitful pattern of moments of engineering creativity, represented by such people as Cherepanov, Mel'nikov, and Lomonosov, followed by failure to sustain this creativity continued to plague Russian railroads just as it did many other technologies.

3 The Electrical Industry: Failed Inventors of the Nineteenth Century

Russian inventors were the first to illuminate the large European cities of Paris and London, they were the first to use incandescent lightbulbs, and they transmitted radio waves before Marconi. Yet none of these ingenious men was successful in business, and as a result, they are forgotten in the West today. The reasons why they failed were societal, not technical. They ran into political, economic, and legal barriers that made it impossible for them to develop further their ideas in Russia. Moreover, in their attitudes toward business they displayed an innocence that has often characterized Russian scientists and engineers down to the present day.

The starting point for the creation of a new branch of industry.

—George Westinghouse, American inventor and industrialist, describing the Russian inventor Pavel Yablochkov's electric lamp[1]

An impressive group of electrical engineers appeared in the last fifty years of tsarist Russia, some of the most notable being Aleksandr Lodygin, Pavel Yablochkov, and Aleksandr Popov. They were a part of a movement in Russia that led to the establishment of electrical engineering institutions just a few years after the very first in the world.[2] It is often said that the world's first courses of study in electrical engineering were given at MIT in 1882, at Cornell in 1883, and in the same year at the Darmstadt University of Technology in Germany. Russia was not far behind. In 1886 the Technical School of Telegraphic Engineering opened in St. Petersburg, becoming in 1891 the Petersburg Electrotechnical Institute.

Lodygin, Yablochkov, and Popov are not well-known names in the West, but they should be. Yablochkov invented the electric lamps that were the first to illuminate Paris and London and inspired Thomas Edison to begin serious research on a lightbulb. Lodygin built the first incandescent lightbulb using a tungsten filament, still used today, a model of which was shown to Edison in 1878, before the latter completed his own lamp.[3] Popov transmitted radio signals before Marconi and was the first person to use an antenna on a radio, with which an early rescue operation saving hundreds of lives was performed.

Just who "invented" electric lights and the radio can be justifiably debated; there are literally dozens of candidates. Two recent authors listed over twenty predecessors to Edison working on incandescent lamps.[4] To the extent that Yablochkov's, Lodygin's, and Popov's names show up in non-Russian sources the question being debated is precisely this question of priority: who actually invented electrical illumination or the radio? Russian nationalistic claims for the priority of their native inventors, especially strongly voiced in the Soviet period, have not helped. The Soviet Union issued postage stamps with the portraits of Yablochkov, Lodygin, and Popov, accompanied by claims that they were, respectively, the inventors of the first city lights, incandescent lightbulbs, and radios. Naturally, those claims evoked resistance elsewhere, and the eminence of Edison and Marconi has been successfully sustained, if public recognition is to be the judge. But the question of priority can be endlessly debated and often becomes semantic. It revolves around what "invention" really means. Does it mean who first had the idea? Or who first demonstrated the idea in a laboratory? Or who first did it in a way that attracted public attention? Or who first made the idea a commercial success?

Distracted by the false allure of the priority question, most observers of the history of technology have missed a far more significant issue about the fates of nations. Why, if Russia possessed such talented electrical engineers at this time, did it not become a leader in electrical technology, and why did it not permanently sustain that success? The special and puzzling place of Russia here is illustrated by the fact that the laurels Edison and Marconi won did not mean that only their native lands, the United States and Italy, excelled in the aborning electrical industry. The creation of the electrical industry was not a zero-sum game in which the countries that first successfully developed innovations reaped all the rewards. All advanced industrial countries participated in that upsurge. But the striking fact is that Russia's place in this great electrical expansion at the end of the nineteenth century and the beginning of the twentieth was minimal, despite the achievements of its engineers. No leading Russian electrical company emerged onto the world stage. Why was that so? When we try to answer this question, we find a remarkable commonality in the lives of Yablochkov, Lodygin, and Popov. All ran into an array of obstacles specific to their homeland—economic, political, and attitudinal. Let us look briefly at their lives and these obstacles.

Aleksandr Lodygin

Aleksandr Lodygin (1847–1923) was born in Tambov province to a very distinguished noble family in terms of ancient roots, although not a wealthy one. This lineage meant he would not have fared well after the Russian Revolution, which he lived to see. Despite his noble background, however, Lodygin, like many "repentant nobles," was sympathetic to critics of the tsarist regime.

Lodygin attended first a cadet school, as was often the tradition among the nobility, then served several years in the army, including at the famed Tula arms works, after which he attended lectures at a St. Petersburg institute of technology. In 1872, at age twenty-five, he applied for an "invention privilege" (not the same as a patent) for an electric incandescent lamp using, at first, a very thin carbon rod as a filament. This was several years before Thomas Edison began research on such a lamp but some years after several other people in France, the United States, and England had demonstrated working but impractical incandescent lights. Lodygin's vacuum bulb worked fairly well, with a filament that lasted longer than some of the others, and he obtained regular patents for it in Austria, Britain, France, and Belgium.

In 1874 the Russian Academy of Sciences awarded Lodygin its Lomonosov Prize for his invention of the filament lamp. That same year, with almost no money, Lodygin launched his own company in St. Petersburg, the A. N. Lodygin Electric Lighting Company. At about the same time Lodygin became very interested in the socialist ideas of the *Narodniki*, or populists, who idealized the peasant commune as the embryo of socialism and who opposed both the monarchy and the development of what it considered exploitative capitalism. Many of Lodygin's friends were in this radical movement. But at the same time that Lodygin embraced these ideas, he was starting his own company and trying to attract shareholders. Was Lodygin a capitalist, with his new company, or a socialist, with his radical friends?

Unfortunately, as even his admiring biographer Liudmila Zhukova wrote, Lodygin "did not know anything about finances and did not want to know anything," he just wanted to be "left in peace" to work on his inventions.[5] This characteristic is one that is observable among many Russian inventors and scientists and is a result of their belief that business is "dirty," even if sometimes one has to engage in it. This attitude is not promising for a person who wants to be a successful entrepreneur. Lodygin was critical of the nascent capitalist environment in which his company had to live. Here was the single biggest difference between Lodygin and his rival Edison, for the latter paid great attention to finances, had many friends on Wall Street, and was deeply interested in creating profitable businesses. In fact, Edison turned on his first electrical grid system, the Pearl Street Station, while standing in the Wall Street office of the financial mogul J. P. Morgan.

Noting Lodygin's aversion to administrative and financial details, several powerful shareholders in his company began to take over the firm and move its assets into their own hands, promising him "freedom" to work as an inventor. They ran the company into the ground, and it ended up in bankruptcy and confusion. The new managers absconded with whatever equipment was left in the factory, and the firm disappeared. Russian biographers of Lodygin have claimed that he had priority over Edison and have thrown the blame for the failure of his company on the managers, but Lodygin, with his hostility to the investment world and his administrative ineptness, deserves his share. Furthermore, beyond the question of personal blame looms the much larger problem of whether an embryonic technology company could succeed in Russia in the face of growing competition from large firms in Western Europe. Russia at this time had little industrial capital and did not have experience in investment and management. Lodygin's personality and his administrative inabilities made an already difficult situation much worse.

An illustration of Lodygin's discomfort with the business world came immediately after the failure of his company. He made a radical shift in lifestyle and joined the "movement to the people" popular among Russian intellectuals at the time, an effort to unite with peasants and rescue a world of rural socialism. He journeyed to the Caucasus, where he joined a commune and worked as a fisherman in the Black Sea, and also as a small-scale inventor helping other members of the commune in their agricultural and marine tasks. This effort soon failed also, and Lodygin returned to St. Petersburg, where he got a job as a mechanic in a factory of Pavel Yablochkov, who, just like Lodygin a few years earlier, was establishing his own electrical company, Yablochkov-Inventor and Co.

Meanwhile, the tsarist government was persecuting radicals, of whatever persuasion. The suppressions grew much worse after the assassination of Tsar Alexander II in 1881 by a man who was allied with the most radical wing, the People's Will, of the populist movement. Lodygin himself abhorred such violence, but he watched as the tsarist police arrested a number of his friends and seemed to be moving closer to him. He responded by leaving Russia and moving to the United States, where he accepted an invitation to work at Westinghouse.

Lodygin remained at Westinghouse only a few years, then moved back and forth between United States, France, and Russia, in each place failing to create successful companies or find a secure position for himself. For a while he worked as an electrician for the New York subway system. In 1908 he sold his patent for a tungsten filament lightbulb to General Electric, a descendant of his rival Edison's original company and a competitor to his old employer, Westinghouse. In Russia in the first years of the twentieth century he lived on very little money, moving from apartment to apartment in search of a less expensive residence. He greeted the first of the two revolutions that struck Russia in 1917, but fled back to the United States when the Bolsheviks became victorious.

The new Soviet government invited Lodygin to return to his homeland to participate in its massive electrification program, called GOELRO. Lodygin may have had a radical past, but he disliked the authoritarianism of communist rule. According to members of his family, his sympathies had been with Alexander Kerensky, the democratic socialist who emerged as a leader in Russia just before the Bolsheviks took over and who, like Lodygin, was forced to flee the country.

Looking back at Lodygin's life we see that he had great gifts as a bench inventor but almost no ability as an entrepreneur. He disdained business, saying it had "no relationship whatever with research talent." In that respect Lodygin symbolized Russia's greatest weakness, its inability to commercialize the ideas that some of its brightest researchers produced.

Pavel Yablochkov

Pavel Yablochkov was born in 1847 to a noble family in the Saratov region of Russia.[6] His family had in earlier generations been wealthy but by the time of Pavel's birth had lost most of its money through division of land among descendants and because of the decline of the rural economy. After a few years of education in a local gymnasium Yablochkov's parents enrolled him in a military-technical school in St. Petersburg because of his talents and interests in technical subjects and because the school was free for noble families preparing their sons for military service. Yablochkov, however, was not destined for a military career; he suffered from poor health, and saw military service as too restrictive for his growing interest in the new field of electrical technology. He obtained a medical release from the army in 1872, at age twenty-five, and soon found employment as a specialist in telegraphy for the Moscow-Kursk railway. This railway had a small workshop in Moscow for the repair and maintenance of telegraphic equipment, which became Yablochkov's first laboratory. And in Moscow he fell in with a group of Russia's first electrical engineers, young men avidly promoting their field by organizing public exhibitions, discussion groups, and, eventually, a museum. Interest in electricity was rapidly growing in Europe, and Yablochkov followed developments there. In 1871 the Belgian inventor Zénobe Gramme demonstrated to the Academy of Sciences in Paris a successful generator of electricity, a dynamo. Soon thereafter Yablochkov and his friends built a similar device in Moscow.

These men began experimenting with batteries, dynamos, and electric arcs and soon built primitive arc lights, which they demonstrated to the public. Yablochkov found an opportunity to gain publicity for himself and one of his new gadgets when Tsar Alexander II in 1874 went on vacation to the Crimea on the Moscow-Kursk Railway. Yablochkov contrived an arc headlight for the locomotive of the tsar's train for use at night. The light was cantankerous and required constant attention, so Yablochkov sat in the open at the front of the locomotive in cold weather tending the light as the tsar journeyed southward. This event was the first time in the world that a locomotive was equipped with an electric headlight.

Though Yablochkov impressed the tsar, his superiors in the railway administration displayed no interest in pursuing electric lights for trains, and the project was dropped. Eight years later French railway engineers

developed such a light and announced it as "the first in the world."[7] Here
was a premonition for later developments in Yablochkov's career; in Russia
there was little interest in business circles in electrical technology. Yablochkov's locomotive headlight was an example of "presentation technology"
(for the tsar) and not a part of the organic development of electrical engineering in Russia.

In Moscow, Yablochkov and a small group of electrical technicians set up
a workshop for the production and repair of electrical equipment, including batteries, dynamos, and arc lights. They offered their services to factories, stores, shipping companies, and railways. No one, it seems, offered
them any business. The electrical workshop failed miserably, running up
so many debts that Yablochkov was threatened with legal action. In 1875
he fled in haste to France, leaving many debts and angry creditors behind,
along with an unhappy wife and children. If he returned to his homeland,
he would face prison.

After an initial period of poverty in Paris, Yablochkov found his way
as a skilled electrical engineer. A local electrical entrepreneur there gave
him a job in his little company, where Yablochkov was able to reproduce
the electromagnets and arc lights he had made in Moscow. In 1875 and
1876 he received French patents for his inventions, including an arc light
that came to be known as the "Yablochkov candle." Between 1876 and
1879, Yablochkov filed six more patents for improvement of his arc light.
In 1876 he created a sensation at an exhibition of scientific instruments in
London, where he headed the delegation for his French employer. He displayed an arc light that dazzled onlookers with its brilliance and effectiveness. Yablochkov's reputation was made.

During the next two years Yablochkov became famous and rich. Together
with his French employer and other investors, Yablochkov launched a new
electrical company. His lights were used to illuminate the centers of Paris
and of London. The first public display of his system was in October 1877,
when he illuminated the Magasins du Louvre in Paris. At the Paris Exposition of 1878 he lighted up a half-mile length of avenue de l'Opera. Later he
illuminated the Paris Hippodrome and the Thames embankment in London. Large stores and luxury hotels in the two cities installed his lights by
the hundreds. Paris was so brightened by his arc lights that it became known
as "the city of light," an appellation still common today. One of France's
largest ports, Le Havre, was equipped with Yablochkov's lights to enable

the nighttime loading and unloading of ships. His lights were being used in many cities, including Marseilles and Rome. Back in Russia Yablochkov was hailed as a hero, and in 1878 his lights were imported from Paris to St. Petersburg for celebrations at the Winter Palace and several other places.

Many countries sent delegations to the 1878 Paris Exposition, including Russia. Yablochkov's lights were a sensation. The American inventor and industrialist George Westinghouse called Yablochkov's lights "the starting point for the creation of a new branch of industry." The Russian delegation was headed by a navy admiral, K. N. Romanov, who also happened to be the brother of Tsar Alexander II, who remembered Yablochkov's headlight for his locomotive in 1874. Admiral Romanov urged Yablochkov to return to Russia and promised him large orders there for his lights to equip the ships, docks, and installations of the Russian navy.

Yablochkov was tempted to return to his native land, where his family and many friends resided, but he had a problem: he was still in trouble with Russian authorities for unpaid debts to Moscow creditors. The way he chose to deal with this problem reveals his naiveté about politics.

In Paris, Yablochkov had become friends with many members of the Russian émigré community, quite a few of them living there in political exile because of past scrapes with the tsarist authorities. One of them was German (Herman) Lopatin, a revolutionary and writer.[8] Lopatin had been imprisoned in Siberia for his radical activities but had escaped to Western Europe. He became friends with Karl Marx and Friedrich Engels, and translated part of Marx's *Das Kapital* into Russian. He was one of the first Russian revolutionaries to become a Marxist. He was skilled at intrigue and made a number of clandestine trips back to Russia, working for radical causes. He offered to take money from Yablochkov, travel to Moscow under a pseudonym, and pay off all of Yablochkov's debts so that he could freely return to his homeland. Yablochkov accepted, and Lopatin successfully carried out his mission. But all this was later noticed by the tsarist secret police. Henceforth, in the files of the secret police Yablochkov was associated with the Marxist revolutionary German Lopatin, although there is no actual evidence that Yablochkov himself was a radical. He just wanted to have his debts paid off, as they were.

Yablochkov returned to St. Petersburg in triumph, living at first in the finest hotel in the country, the European Hotel (today called the Hotel Grand Europe). He was toasted by all of society and rapidly ran through his money returning the hospitality. To buy from his co-investors in France the rights to manufacture his lamps in Russia he had to pay them a million francs. He established a new company in St. Petersburg and offered his lamps for sale. The Russian navy was true to its word and was his best customer, buying several hundred lamps. But Russian hotels, businesses, and factories displayed remarkably little interest in Yablochkov's lamps. The cities of St. Petersburg and Moscow would not be electrified for many more years, and when electrification finally came, the city authorities contracted with the German company of Siemens and Halske to do the job. By then Yablochkov was no longer a competitor.

Yablochkov once again went bankrupt and once again returned to France, hoping there to repeat his earlier triumphs. However, the field of electrical illumination was changing in a way that Yablochkov had not adequately noticed. His arc lamps emitted a harsh, glaring light that was not suitable for individual use in homes, though they were fine for public spaces. Thomas Edison knew that the real future of electrical illumination was in smaller lights with softer illumination, of the sort that a person reading a book might wish to use. Even though Edison was not the inventor of the incandescent light, he was unparalleled in his ability to improve it and, even more, to make it part of an electrical grid system with generating plants that would supply entire cities, including private homes. And Edison worked closely with Wall Street financiers.

Yablochkov lost out in this competition. Never again did he achieve the success of 1875–1878 in Paris, although he tried again and again, both in France and in Russia. He died in Russia in 1894, a broken and sick man. One reason why he had difficulty succeeding in Russia was that the tsarist authorities, except for the navy, were suspicious of him. The tsar's government feared the political independence that successful entrepreneurs would inevitably attain, and Yablochkov's association with radical revolutionaries such as Lopatin increased their anxiety. And businesses in Russia were not interested in technological innovation. Even the luxury hotel where Yablochkov lived for a while in St. Petersburg would not buy his lamps, preferring traditional gas lanterns.

Yablochkov was an outstanding inventor but a poor businessman. He worked essentially alone on his inventions, never having the advantage of a whole research laboratory of the sort created by Edison, whose greatest invention of all was probably "the invention of invention." Russian biographers of Yablochkov have praised him in comparison to Edison, whom they have often seen as opportunistic, working not for the sake of knowledge, but just taking advantage of whatever society wanted at the time, whether it be a phonograph or a lightbulb, and at the same time borrowing the ideas of other inventors, such as Yablochkov and Lodygin. But that is just the point that Russia needed to understand most: success in technology is as much a matter of business acumen as it is of technological brilliance. In fact, invention is probably the easy part of commercial success in technology. The hardest part is knowing the market and fitting innovations to economic opportunity. Yablochkov, like many Russian inventors, never mastered this skill, and the society in which he lived did not prize it. Thus, although George Westinghouse said that Yablochkov was the "starting point for a new branch of industry," that industry was not in Russia.

Aleksandr Popov

Aleksandr Popov (1859–1906) was born in the northern Urals region on the border of Russia and Siberia. The son of a priest, he was a member of a family with a strong tradition of education, and several of his brothers and sisters also went on to higher education. The Urals area was the home of mining and industry, and from an early age Aleksandr was surrounded by, and engaged with, mechanical devices. Popov began his education at religious schools, first in the local town of Dolmatovo and then in the city of Perm. He soon established a reputation as an outstanding student.[9] On the basis of competitive examinations he won the right to attend St. Petersburg University, where he had been preceded by his older brother, who gave him much assistance. St. Petersburg University was at this time probably the academically strongest university in Russia. Popov chose as his special area of study the field of physics. On graduating in 1882 he was selected to work as a laboratory assistant at the university in the hopes that a more substantial position would open up for him, but no faculty vacancies appeared. Popov then accepted a position teaching at the Russian navy's school of munitions in Kronshtadt on Kotlin Island in the bay of the Baltic near St. Petersburg. After 1889 he became very interested in the work of Heinrich Hertz, who just a few years earlier had demonstrated the existence of electromagnetic waves.

In 1894 Popov built his first radio receiver, first designed as a lightning detector, and then both a transmitter and receiver. On May 7, 1895, Popov demonstrated to the Russian Physical and Chemical Society the successful transmission and receipt of radio waves. Popov achieved a range of 600 yards in 1895, 6 miles in 1897, and 30 miles in 1898, when he sent a telegram with the words "Heinrich Hertz." He established a radio station on Kotlin Island, and another one, in 1900, on Hogland Island (Suursaari). Soon he was sending radio messages from Russian navy bases to ships.

A dramatic demonstration of the importance and effectiveness of Popov's radios came in 1900 when the battleship *Apraksin* went aground in ice floes in the Gulf of Finland with hundreds of sailors and officers aboard. The battleship remained aground for months, but during that time 440 official telegram messages were sent between it and naval officials ashore by Popov's radio stations. Based on these messages the battleship was supplied by an icebreaker, and the crew was eventually rescued. This event brought Popov great publicity and honor. He had demonstrated the indisputable value of his work to his navy superiors.

Popov was then offered and accepted a position at the Electro-Technical Institute in St. Petersburg and within a few years had advanced to high administrative positions there. In these early years of the twentieth century Russia was a turbulent place politically, with universities often closed down by student strikes and demonstrations. The modest and self-effacing Popov was ill prepared for these stresses. The humiliating defeats suffered by Russia in the Russo-Japanese War inflamed radicals and conservatives alike, and the workers' movement in the capital city became increasingly militant. By early January 1905, St. Petersburg was paralyzed by strikes, and had no electricity and no newspapers. Then on January 22 (January 9 old style), "Bloody Sunday," armed soldiers at the Winter Palace fired on peaceful demonstrators, killing hundreds. This event led to an explosion of protests and was one of the causes of the 1905 Revolution. Popov, along with many other members of the Russian Physics and Chemical Society, signed a letter of protest. Newly elected to be the director of the Electro-Technical Institute but not yet actually in office, Popov was the natural place of appeal by faculty and students alike. He came under enormous pressure, which he was not constitutionally able to withstand. He fell ill, and on December 31, 1905, at the age of forty-six, he died of a brain hemorrhage.

From that time to the present day Popov has been a hero in Russia, where he is almost universally hailed as "the inventor of the radio" and is favored overwhelmingly over the man most frequently given the award abroad, Guglielmo Marconi. And Popov has an absolutely legitimate claim to be one of the handful of pioneers in the development of the radio. Other strong candidates in addition to Marconi are Nikola Tesla and Oliver Lodge. When we look at the dates for crucial experiments by these four men in the invention of the radio we see that they are separated by only months: Nikola Tesla (1893), Oliver Lodge (1894), Aleksandr Popov (1895), and Guglielmo Marconi (1895). These men, operating in four different countries—the United States, Britain, Russia, and Italy—were all thinking about wireless telegraphy at about the same time. The question of who built the first radio, to which dozens of books have been devoted, is not the most important question. Much more meaningful than the issue of technical priority is one of human significance and benefit: who first made radio a successful commercial product that changed the lives of millions of people? And this query leads directly to another, why was only one of these men—Marconi—a great commercial success? And the answer to this latter question reveals, among other things, why the Russian candidate was not a success, and why most people outside Russia even today have never heard of him.

Popov considered himself a scientist, not an inventor seeking a commercial project. He did not take out patents early enough to protect himself. He was simply not interested in such an effort. He thought he had established priority by giving papers at scientific societies, not realizing there were people around who would soon make his ideas their own. Popov was a typical "Russian intellectual," actually proud of the fact that he had no commercial interests. His major Russian biographer asserted that Popov was the "embodiment of the best characteristics of the Russian *intelligent* . . . of modesty and indifference to wealth, and of care only for the good of the people."[10] Popov was obviously an admirable man. But another question begs to be asked: does it really serve the "good of the people" to have important ideas that are not used for their benefit?

There are elements of a morality play in the contest between Marconi and Popov, still raging in Russia. Marconi was the antipode of the simple Popov: the son of a wealthy Italian landowner, a skilled businessman, socially adept, a master of publicity, an unabashed opportunist, an acquisitive capitalist, and the manager of a financial and industrial empire. He became a multimillionaire and lived in the last part of his life on a marvelous white yacht with a crew of thirty and a laboratory where he conducted radio experiments. He circulated in the highest levels of society and entertained kings and queens on his yacht, as well as multiple mistresses.[11] Newspapers all over the world celebrated him as the man who brought the radio to humanity. Popov, on the other hand, was entirely alien to business interests and, to the extent that he was interested in his place in history, it was based on his scientific reports. Marconi bragged that he was not a scientist, just a clever inventor, and he outmaneuvered his competitors with multiple patents, many of them disputable. Marconi was even accused of stealing the ideas of others, and lost a number of patent cases. But he became a celebrity, and descendants of his companies still exist today. Few people outside Russia have heard of Popov, and no Russian company today can claim direct descent from him. Once again, Russians were good at ideas, poor at business.

Popov is still highly honored in Russia today. St. Petersburg Electro-Technical University is located on Popov Street. The handout literature of the university describes Popov as its first director and the inventor of the first "wireless receiver." May 7 is celebrated as Radio Day throughout Russia, the date being the anniversary of Popov's 1895 successful demonstration of radio transmission—earlier, Russians say, than the "upstart" Marconi. The Russian enmity toward Marconi is further spurred by his politics. Marconi became a leading fascist in Italy, and Benito Mussolini was best man at his second marriage.

4 Aviation: A Frustrated Master, a Deformed Industry

Russian aircraft designers and aeronautical engineers have displayed great creativity from the beginning of the air age. One of them designed and flew four-engine passenger planes, complete with lounges and washrooms, only a few years after the Wright brothers first flew in 1903. Others designed planes in the 1930s that established sixty-two world records, including the longest, highest, and fastest flights in the world at that time. Yet the Russian aviation industry was deformed by political requirements and never produced planes that could successfully compete commercially with Western rivals. As a result, Russian airlines today are increasingly using Western airplanes produced by Boeing and Airbus.

I doubt if this could have taken place anywhere else in the world.

—Igor Sikorsky, aviation pioneer, describing his successful work in the United States after failing to be able to develop further the remarkable planes he first constructed in Russia

Igor Sikorsky (1889–1972) was a creative aviation pioneer whose life demonstrates the great promise that Russian technology has long held but that has not been fulfilled because of the social, political, and economic conditions of Russia.[1] Like many of the great technological innovators of the twentieth century he was originally an outsider, a college dropout who relied on his own resources to promote his inventions. Borrowing money from his family of modest means, in the first years of the twentieth century, just a few years after the Wright brothers' first flight in 1903, Sikorsky built in his father's garden in Kiev a series of airplanes cobbled together from bedsteads, bicycles, piano wire, and hardware purchased in junkyards.

The authorities of the day understandably dismissed these contraptions as unflyable, and several of Sikorsky's early attempts were indeed dramatic failures. Furthermore, the tsarist police were suspicious of flying machines, fearing they would be used for subversive purposes, and actually established a special commission for "Battling the Possible Implementation of Criminal Designs With the Assistance of Aeronautical Machines."[2] Nonetheless, Sikorsky persevered and eventually built the S-6, a single-engine biplane with an enormous and graceful curved propeller that required hand cranking to start.

The Russian Army announced a competition for the best airplane in which the criteria for the prize were speed, length of take-off, landing speed, length of landing, lifting capacity, and ability to land and take off from a plowed field. Sikorsky got in his plane and started down the plowed field at a carefully chosen moment, a cold morning when frost made the furrows dauntingly rough but sufficiently hard to support the wheels of his aircraft. He successfully took off and won the prize. He was twenty-three years old.[3]

Sikorsky was then offered a position to head up the division on aviation in the Russo-Baltic Train Car Company, a large firm near St. Petersburg. With the newly acquired salary that his position brought he paid off with interest the loans he had taken from his father, a professor, and his sister, who ran a home for disadvantaged children. The family finances, battered by Sikorsky's airplane expenditures, recovered.

Being inducted into the Russian industrial and governmental establishment brought disadvantages as well as advantages. Sikorsky wanted to build passenger airliners that would open up commercial aviation, and he soon constructed and flew from St. Petersburg to his home city of Kiev the world's first multiengine airliner. His plane was a four-motor giant that could carry sixteen people, a world record at the time. The plane attracted much attention, including that of Tsar Nicholas II himself, who climbed a rickety ladder into the balcony of the airliner to inspect it personally. The plane included a salon, dining table, washroom, and comfortable wicker chairs.

The Russian government and Russian business community were not ready for such an innovation. Instead, they wanted a military bomber that could be used in warfare. Sikorsky complied, replaced the salon with a bomb rack, and built more than seventy four-engine strategic bombers, which saw much action in World War I. (Not many people today are aware that four-engine bombers existed in World War I; such planes do not fit into romantic legends of the Red Baron and of combat among modern knights in open cockpits in single-engine biplanes.) Sikorsky's bombers had large enclosed cabins and were the first such planes in history.

When the Germans saw this new threat, they attacked the bombers from the rear, an angle of attack that was at first very frightening to Sikorsky's pilots. Sikorsky then installed rear turrets on his bombers, manned by gunners, who shot down a number of German planes. Only one of his bombers was lost in the war, even though they carried out hundreds of raids and were especially effective against railroads bringing supplies to the German troops.

When the war ended, Sikorsky hoped to return to his dream of commercial aviation. However, the Russian Revolution intervened. His employer, the director of the Russo-Baltic Train Car Company, was shot by the revolutionaries, and the firm was taken over by the new Soviet government, which declared its opposition to private enterprise. Several of Sikorsky's employees, including his best test pilot, were also executed. Sikorsky was a deeply religious man and abhorred the Soviet attitude toward religion. To the Soviet authorities Sikorsky was an unreformable "bourgeois specialist," opposed to the new order.

Sikorsky's attempt to find support in the new Soviet government for his dream of commercial aviation was unsuccessful. Just like their tsarist predecessors, the Soviet authorities wanted military planes. Despite his intense Russian patriotism, Sikorsky decided to leave his native land, going first to France, where many aviation pioneers also worked. But in France he could not find the opportunities he needed, and in 1919 he moved to the United States.

Fulfilling his dream of commercial aviation was not easy in the United States either. He started out building planes on the farm of a friend, then found some support from private investors, including the Russian composer Sergei Rachmaninoff, who provided the single most important early financial commitment. Eventually Sikorsky created his own company, Sikorsky Manufacturing Company, which became a part of United Aircraft Corporation and still exists today as Sikorsky Aircraft Corporation, a division of United Technologies Corporation. Sikorsky's name in the United States became almost synonymous with helicopters, but he was also a leading airplane designer.

Sikorsky's dream of commercial aviation was realized when he built the Pan American Clippers, large amphibious planes with which Pan American Airlines opened up the South Pacific and other far-distant areas to commercial aviation in the 1930s. These planes could carry over thirty passengers and, for overnight flights, offered fourteen sleeper berths with a lounge. Sikorsky got his salon back after all.

This short history of early aviation in Russia shows that the obstacles Sikorsky met were primarily social, political, and economic, not technological. One is tempted to ask, what would have happened to Sikorsky and his dreams if the Russian Revolution had not occurred and he had stayed in Russia? It would be easy to think that this talented engineer would have succeeded there as he did in the United States. However, this conclusion is unwarranted. Neither the tsarist government nor the Soviet one favored individual initiative and private enterprise, and neither provided an environment in which private investors could support talented innovators. Even in France, where democratic government and freedom were basic principles, Sikorsky could not obtain the support he needed. Of the United States he later wrote in his memoirs, "I am deeply grateful to this great country of unequaled opportunities which enabled me to resume my life's work. . . . I doubt if this could have taken place anywhere else in the world." A better example of the need for technology to have a supportive social and economic environment can hardly be found than in the life of Sikorsky.

Of course, the Soviet Union established its own aviation industry, one that was in some ways impressive. However, Soviet leaders were more interested in using aviation to legitimate their own rule and to demonstrate superiority over Western countries by setting speed, endurance, and distance records than they were in developing planes that were efficient and economically competitive in world markets. As a result, the Soviet Union developed an aviation industry that was malformed.

Stalin in particular wanted his "falcons," as he called his pilots, to impress the rest of the world and to demonstrate the superiority of the Soviet socialist system. In the late 1920s and during the 1930s he called for them to fly "Higher, Faster, Farther!" In 1929 he dispatched the newly developed Tupolev ANT-9 monoplane, the prototype of which was named *Wings of the Soviets*, on a 5,600-mile European tour. A similar mission was given to the Tupolev ANT-4 plane, called *Land of the Soviets*, which was launched in the same year on a 13,000-mile flight to the United States.[4] Stalin's desire for spectacular flights meant that he wanted giant airplanes, a demand that resulted in the construction of the ANT-20, an eight-engine monster that had almost no virtue other than size. Dubbed the *Maxim Gorky*, the gigantic plane was shown at public demonstrations in Red Square and elsewhere. Unfortunately, on May 18, 1935, while showing off to admiring crowds, it was involved in a collision with a smaller plane that killed forty-eight people.

Undeterred, Stalin summoned aviators to the Kremlin and charged them with making spectacular transpolar crossings. One of the most famous was piloted by Valerii Chkalov in 1937 and flew from Moscow via the North Pole to Vancouver, Washington. His plane, an ANT-25, was specifically designed for the feat, with an extreme wing aspect ratio that made it almost unusable for other purposes. But Stalin was interested in laurels, not commercial viability, and by 1938 he could claim sixty-two world records, including the longest, highest, and fastest flights.[5]

The Soviet aviation industry that developed before World War II bore the imprint of political authoritarianism. If Stalin's plane designers displeased him, he imprisoned them, as he did with two of the most prominent ones, Nikolai Polikarpov and Andrei Tupolev.[6] In his desire for spectacular achievements he distorted the entire industry. To the extent that he had interest in practical aircraft, it was primarily for military uses. As a result, there was little independent creativity, and almost no interest in commercial advantage.

After World War II, and especially after Stalin's death in 1953, Soviet aviation came closer to normality. Nonetheless, the habits of design that had been acquired in the Stalinist period continued to plague Soviet aviation. The least important criteria in Russian aviation design were fuel economy, safety, attractiveness, and comfort. As a result, when Soviet aviation was exposed to international competition after the disappearance of the Soviet Union, Russian manufacturers could not match the products of the foreign aviation industry. Russian airlines today increasingly use planes manufactured outside Russia by companies such as Boeing and Airbus. However, the Russian airplane company Sukhoi, known best for its supersonic fighter jets, is currently making a major effort to enter the international regional jet passenger market with its newly designed Superjet, which seats one hundred people. The Superjet has attracted international attention, but so far sales have been modest. The reputation of Sukhoi has not been helped by the revelation in 2010 that seventy of its engineers had bribed a local technical college to give them fake engineering diplomas.[7] That reputation was further depressed by the crash of a Superjet with several dozen passengers during a demonstration flight in Indonesia on May 9, 2012.

5 Soviet Industrialization: The Myth That It Was Modernization

Many people believe that Soviet industrialization was a success. After all, a largely agrarian country was transformed in a few decades into an industrialized one, and the Soviet Union succeeded in halting Hitler's armies in World War II. However, this achievement was warped and distorted by political edict in such a way that Russian industry today is noncompetitive with its international rivals. Russian factories were often built in the wrong place, in the wrong way. Political and ideological considerations trumped sound engineering and economic ones. As a result, the Soviet industrial system is for Russia today as much an impediment as it is an advantage.

Russia is not an underdeveloped country; it is a badly developed one.
—Thomas W. Simons, Jr., American diplomat in the Soviet Union, former ambassador to Poland and Pakistan and lecturer in government, Harvard University

Although the Soviet Union was a failure as a political and economic system, many people even today consider its industrialization and modernization program a success. They see that a predominantly peasant and agrarian nation was transformed into an industrial power of world standing. And, after all, the Red Army was able to match and overcome the technological might of Hitler's armies. The Soviet Union built up an industrial system that made it at one point the second largest economy in the world. On Soviet soil during the Five-Year Plans launched before World War II there arose the world's largest steel mills and largest hydroelectric power plants. Foreign observers and participants, from the photographer Margaret Bourke-White to the labor leader Walter Reuther, came to witness and admire the "great Soviet experiment."

But was this modernization? As time goes on, we are able to see more and more clearly that it was not, certainly not if modernization includes the application of rational analysis to the solution of social and economic problems. In fact, present-day Russia is saddled with an industrial system and a population distribution that were so distorted by central fiat during the Soviet period that they are major obstacles to true modernization today.

Factories, power stations, canals, railroads, cities, and enormous complexes were built in the wrong places and in the wrong ways, if evaluated by cost-effectiveness. The construction of industry was governed more by military and ideological considerations than by economic ones. This "legacy of the past," as several authors have called it, makes Russia's further economic development extremely difficult.[1] In a globalized world of international competition in the sale of advanced technologies, Russia's present inefficient production infrastructure puts it at a signal disadvantage. Rather than building on Soviet industrialization, contemporary Russia needs to overcome it. Examining the ways in which the Soviet Union industrialized will aid in understanding today's situation. As we will see, industrialization could have been done much more efficiently and effectively but was thrown off course by political considerations. Certain early Soviet industrialists and, especially, engineers tried to warn Soviet leaders of the mistakes they were about to make. Unfortunately, they were not heeded.

The triumph of the communists in 1917 (in the second of the governmental overturns of that year) was met with hostility by the great majority of scientists and engineers in Russia. Some emigrated, while others remained, but most of those who stayed in place fell silent politically, hoping to continue their technical work without interference.[2] Maybe they thought the radicals who had taken over the government would quickly fail because of their own inadequacies, and this unfortunate episode in Russian history would pass. As the months dragged on, however, and as it became ever clearer that the new Soviet government was going to stay, and that one of its goals was to industrialize and modernize Russia, an important segment of the engineers, those most interested in industrial planning, began to think it might be possible to work with "the Bolsheviks" after all. The engineers were committed to rational, "scientific" analysis, and they were not opposed to the idea of central planning so long as it accorded with their conceptions of sound and efficient policy.

Among the best known of these engineers educated before the Revolution who began to work with the Soviet government were I. A. Kalinnikov, Iu. V. Lomonosov, P. K. Engelmeier, R. E. Klasson, L. Ramzin, N. F. Charnovskii, S. D. Shein, V. I. Ochkin, and P. A. Palchinsky. Perhaps the most outspoken was Peter Palchinsky, a talented and energetic mining engineer who was convinced that the mineral wealth of Russia destined the country for industrial greatness if only rational and fair policies for the exploitation of this wealth were drawn up and pursued by the government.[3] He was one of the best-known engineers in early Soviet Russia, serving as chairman of the Russian Technical Society and a member of the governing presidium of the All-Russian Association of Engineers. He was constantly consulted by the government, wrote many reports for government commissions, and everywhere championed the cause of industry. He advised the government on projects such as the building of a giant dam on the Dnieper River, the mining of iron deposits in the Urals region, and the building of seaports, canals, and railroads.

Central to Palchinsky's approach to planning industry was the concept of "achieving the greatest possible useful effect with the least waste of human and financial resources."[4] This approach meant that before any large industrial project was launched, all the available alternatives and variants should be carefully studied to see which would be most efficient. These variants included not only the different technologies that were possible for a given goal but even different physical locations for factories, with the one ultimately chosen being most efficient in terms of access to transportation, a critical population of workers, and the costs of heating and power.

Palchinsky believed that the traditional engineering curriculum was too heavily weighted toward natural science, mathematics, and "descriptive technology" and almost totally ignored subjects such as economics and political economy.[5] In tsarist times the officials kept such subjects out of the curriculum because they feared Western "radical" economic and political ideas. In the new Soviet period Palchinsky feared that different ideological concerns would have similar effects. He called for Russian engineers to stop approaching problems in a narrow technocratic way and to evaluate industrial projects in all their aspects, particularly the economic ones. And he believed that concern for satisfying the needs of workers was not just an ethical principle but a requirement for efficient production. He stressed that successful industrialization and high productivity were not possible without highly trained workers and adequate provision for their social and economic needs. An investment in education promoted industrialization more than an equivalent investment in technical equipment since an uneducated or unhappy worker would soon make the equipment useless.[6] Simply adding new equipment without attention to the morale and ability of the workers would cause great waste.

When the Soviet government proposed building one of the world's largest hydroelectric power plants on the Dnieper River,[7] Palchinsky and his engineering colleagues (especially R. E. Klasson, a specialist in electric power) approached the project in their usual methodical way. They agreed that the Soviet Union needed much more electrical power and were enthusiastic about building more generating plants, but they wondered whether building such a large dam in that particular place was the best way to achieve the goal. They noted that there were plentiful coal supplies nearby and observed that the decision whether to build a hydroelectric power plant or a thermal one should be based on calculating social and economic costs. They maintained that a thermal plant would be needed in any event, since the level of water in the Dnieper River was inadequate to generate much electricity from December to February. They also noted that the proposed dam was in flat topography, on the floodplain of a river, unlike most hydroelectric power plants in other countries (which were usually located in deep valleys or canyons), and that a large number of people and a great area of land would be affected by the resulting reservoir.

To clear the area for the reservoir, more than ten thousand villagers would be forced out of their homes. Most of them were German Mennonite farmers, prosperous and industrious. The critics noted that the loss of these farmlands and the diminished food production that would ensue if the farmers were not allowed to continue in their present location should be included in the expenses of building the dam and reservoir.

Palchinsky admonished the government not to build enormous hydroelectric plants such as the one being planned for the Dnieper River without taking into consideration the distance between where the energy would be used and where it would be generated. Long-distance transmission lines, he warned, involve huge transmission costs and declines in efficiency. At each more distant point along a long transmission line the costs of the delivered energy rise, and the chances increase that less expensive electricity could be obtained from a local source.

Taking all these factors into consideration, Palchinsky, Klasson, and other engineers recommended starting with the construction of one or two steam plants at the Dnieper site and then expanding step by step in line with the demand for power in the region, combining hydroelectric and thermal plants as needed.[8] At more distant locations from the site other forms of power generation should be considered, with the most economical variant chosen.

When the Soviet government proposed building the world's largest steel plant in Magnitogorsk, Palchinsky and his fellow engineers raised similar concerns. The leaders of the Soviet government suggested that the steel plant be located in Magnitogorsk because it was the site of one of the country's richest iron deposits, known as Magnetic Mountain.[9] To nonspecialists such as Communist Party leaders, this location seemed sensible. But was this really the best place for such a large steel mill? In articles published in 1926 and 1927 Palchinsky complained that the Soviet government was going ahead with these plans without adequate studies of the amount of iron ore deposits in Magnitogorsk, the quality of the ore, the availability of labor, the economics of transporting the ore, and the difficulty of supplying proper housing for the workforce.

He noted that no coal was available near the projected city of Magnitogorsk, so that from the very beginning fuel for the voracious blast furnaces would have to be hauled in by rail. Since nobody knew just how much iron ore was there, it was possible that in a relatively short time both iron ore and coal would have to be hauled from far away to the misplaced steel city (this is exactly what later happened).[10]

Knowing how other countries planned the siting of their steel mills (he kept up with the literature in English, German, Italian, and French), Palchinsky pointed out that putting the Soviet Union's largest steel mill in Magnitogorsk would be counter to "best practices" elsewhere. In the United States, he noted, the steel mills were not located near the rich iron ore deposits of the Mesabi Range in Minnesota or the Marquette Range in Michigan but hundreds of miles away, in Detroit, Gary, Cleveland, and Pittsburgh—all cities with large labor forces, the first three connected to the sources of ore by water and the last located near enormous coal deposits. The selection of a site for an industrial facility, Palchinsky emphasized, had to be based on multiple factors, no one of which—such as the location of the raw material—was governing. He called for the drawing up of gravimetric charts, magnetrometric measurements, and economic calculations, and for a cost analysis of methods of transportation and freight engineering. The costs of building the city of Magnitogorsk and its mills might be so great, Palchinsky continued, that it would be wiser to expand steel production near less rich deposits of iron ore but at locations with better labor and transportation resources.

A third gigantic project of the early Soviet industrialization program was the White Sea Canal, designed to connect the White and Baltic Seas, a dream that went back to the time of Peter the Great.[11] A team of Russian engineers, headed by a friend of Palchinsky's, N. I. Khrustalev, was asked by the Soviet government to plan the path of the canal. The engineers asked whether it might not be more economical to build a very good railway between the two seas instead of a canal, since a railway would be operable year-round while a canal in that northern region would be frozen half the year. When told that Stalin wanted a canal, not a railway, they responded that the water supply for the canal was a serious problem and that the most direct route, favored by the government, would not supply enough water for large ships, only small ones and barges. They suggested another "western variant" that would be deep and have a secure supply of water. The disadvantage of the western variant was that it would take longer to build and would require more mechanized equipment.

In each of these cases—the building of the gigantic hydroelectric dam on the Dnieper River, the construction of the world's largest steel mill in Magnitogorsk, and the digging of the White Sea Canal—the recommendations of the local Russian engineers were overruled by the Soviet government. That government was not interested in cost-effective analyses; it was devoted to large-scale projects and their revolutionary symbolism, regardless of the local conditions that the engineers found so important. Stalin demanded that industrial establishments be of great size, preferably the largest in the world—an industrial policy that Western observers later characterized as "gigantomania." In contrast, Palchinsky maintained that size was not a virtue by itself. He asked rhetorically, "Is it possible to build locomotives, oceangoing ships, bridges, and gigantic hydraulic presses in small workshops or handicraft centers? Of course not. But do we need gigantic factories to have good buttons, good socks, office materials, tableware, clothing, etc.? Of course not."[12]

The Soviet Union, Palchinsky warned, must have a goal beyond the construction of heavy industry for symbolic and ideological purposes; it must also aspire for a society where all human needs were economically fulfilled. This goal was required, he said, both for the citizens themselves and for the competitive strength of the Soviet Union compared to other countries, which were building industrial capacity in rational and cost-effective ways. The result of the Soviet policies regarding industrialization, he predicted, was an industrialized country that was inefficient and noncompetitive.

The recommendations of engineers like Palchinsky were ignored, and the engineers themselves were accused of "wrecking," of purposely trying to slow down the rate of Soviet industrial expansion by frustrating the suggestions of the party leaders. Palchinsky was executed without trial, and almost all of the engineers who questioned governmental industrial policy were arrested.[13] A new generation of Soviet-educated engineers was fast coming, the largest educated group in the Soviet Union. These were people who had learned not to question government policies. They built factories where and in the way they were told to build them.

As a result of these policies, which were followed for the next sixty years, Russia today has an industrial system that is badly misallocated. Magnitogorsk is today one of the largest steel mills in the world and one of the most inefficient, forced to import all of its major raw materials, coal and iron ore, from far away. The Dnieper Dam still operates today, but, as a Russian environmentalist observed, the money spent to control erosion and algae in the reservoir "has long since exceeded the short-term advantages the power plant once yielded."[14] The White Sea Canal still exists but has such water shortage problems that large ships cannot use it, even though its use by such ships had been one of the government's rationales for its construction. And these are just three examples of a far larger pattern throughout the entire country.

Soviet ideology included a strong command to "conquer nature," which meant building cities and factories in the far north, in arctic regions where they never should have been built.[15] Two Western authors have called this pattern the "Siberian curse."[16] Russia today has cities with millions of residents located in places so cold and so expensive from supply, heating, and energy standpoints that they cannot compete in the world marketplace. Perhaps the most graphic way of illustrating this is by comparing the growth of two northern cities in the United States and Russia with industrial potential: Duluth, Minnesota, and Perm, Russia. Duluth is located near the largest iron deposits in North America and was, in the early twentieth century, predicted by some demographers to become a major American city. Perm is located near the Ural Mountains, where there are large deposits of valuable minerals. Both Duluth and Perm are very cold cities.

At the beginning of the twentieth century both Perm and Duluth had populations of less than 200,000 people. Today Perm has a population of almost a million and Duluth has a population of just over 200,000. In other words, Duluth has not grown much at all, while Perm has exploded. It would be a mistake to think that this difference is a result of the more impressive growth of industry and population in Russia than in the United States. (Russia's population a little more than doubled between 1900 and 1990; the U.S. population a little more than tripled in the same period. The U.S. economy grew about three times as fast as Russia's in the same period.) The reason for Duluth's failure to grow significantly was documented in a classic 1937 case study by two American economic geographers. They concluded that the cold winters in Duluth increased steel production costs; that Duluth had high labor costs because of the cold climate, necessitating a cost-of-living adjustment; and that Duluth was at a competitive disadvantage relative to other American industrial centers owing to its distance from markets. These same considerations would apply to Perm, but they were ignored because of the Soviet ideology of conquering nature, of the Soviet leaders' emphasis on industrializing underdeveloped regions, and of the decision to ignore economic factors in doing so. As a result, Perm is today what the analysts Hill and Gaddy call "one of Russia's frozen dinosaurs, the world's fifth coldest city with a population of over a million people, bereft of the rationale that built it to such proportions in the first place."[17]

Another graphic illustration of the irrational planning of Soviet industrialization was the Baikal-Amur Mainline (BAM), the largest construction project in post–World War II Russia. Tsarist Russia had also often built railways by edict and employed a logic governed more by military and political considerations than by a rational analysis of the transportation needs of the economy, but in building BAM, Soviet Russia pushed this principle to an absurd extreme.

BAM is almost two thousand miles long, includes 21 tunnels and 4,200 bridges, and required several decades to build, by approximately 500,000 workers. It was also an economic blunder of massive proportions. No significant economic benefit has yet come from it. No wonder the Western historian of the effort called it "Brezhnev's folly."[18] It is a monument to the errors that can be made when central authority calls the shots without adequately considering alternatives and conducting cost-benefit analyses.

The rail line stretches from the city of Novokuznetsk to the Pacific Ocean and goes through some of the most difficult terrain on earth, over and through mountain ranges, swamps, and rivers. Its construction required working in the winter in temperatures so low that equipment failed and tools shattered. More than twenty towns and cities were built along the way. As one Soviet book announced, "The Baikal-Amur outranks every other project in the history of railroad building anywhere in the world."[19]

The original promoters of BAM promised the unlocking of Siberia's hidden mineral riches and the flourishing of cities all along its route. Particularly important was access to the rich copper deposits of Udokan, in Chita province. The shipping of Siberian oil, along with coal and timber, to the Pacific Ocean for export was also a key goal. The plan's supporters spoke of creating a "mighty industrial belt along the BAM," including a major metallurgical complex.[20] Less often mentioned, but also a significant motivation for the project, was the wish to construct a railway that was protected from potential seizure by the Chinese in case of conflict. The older Trans-Siberian Railway runs near the Chinese border for hundreds of miles, while BAM was placed farther north, in a more secure area.

Leonid Brezhnev announced the beginning of BAM's construction in 1974. He described it as a continuation of the tradition of "labor accomplishments by our people" such as the Dnieper hydroelectric station and the steel city of Magnitogorsk.[21]

The construction of BAM was the last gasp of the old Soviet methods of organizing work by hortatory celebration, with minimal attention to difficult technical and social issues. The railroad project was organized like a military campaign that had to be successfully completed in a hurry, at any cost. It displayed many of the same profligate characteristics of all large Soviet construction projects since the early thirties. The decision to build the railway was not an open process in which defenders and proponents of the project or its different variations could publicly present their views, with supporting evidence; instead, the decision was made by the leaders of the Communist Party, relying on a small circle of technocratic advisors nurtured on the familiar ethos of extensive development of production facilities without regard for the costs—economic, environmental, or social.

Once the decision to build the line had been made, anyone who voiced criticism was immediately branded a skeptic, a laggard, a person lacking enthusiasm for the construction of communism. Such a person was not sent to prison, as might have been done in the time of Stalin, but was simply denied publication by the government-controlled media. Soviet newspapers, radio, and TV exalted BAM as a breathtaking challenge requiring wartime fervor, perseverance, and engagement. The enemy was not a hostile army but nature itself—the frozen taiga in the winter and permafrost and mosquito-ridden swamps in the summer. In the face of these obstacles, success was achieved through sheer determination. There was no place for carpers and critics.

Despite the exhortations of the government, the BAM project soon fell far behind schedule and grossly exceeded the inadequate cost estimates. The supervisors naturally looked for ways to speed construction and cut costs. One way was to use lighter rails than planned. In the late seventies and early eighties light R50 rails were used on several sections of the track, instead of the hardened and heavier R65 rails.[22] As a result, three train wrecks occurred before the railway was even completed, and the rails quickly began to deteriorate, especially on the curves. Eventually all the light rails were replaced, but only after the supervising engineers falsely informed their superiors that the track had been completed as specified. Safety and the long-term budget were sacrificed to the time schedule and the short-term budget, a typical result of accelerated tempos.

Another way to hasten construction was by using military troops in construction brigades. The employment of soldiers in construction was a traditional practice in the Soviet Union. Soldiers had helped build the Ignalina and Gorky atomic power stations and many of the irrigation canals along the Volga River. Indeed, they could be seen on almost every Moscow street, repairing roads and putting up buildings. Whenever a supervisor experienced difficulty in getting a certain project finished on time he would appeal to the Ministry of Defense, saying, "Unless you give us soldiers, the plan is lost."[23] The builders of BAM were no different. Much of the eastern section of the railway line was built by soldiers, who also did 25 percent of the heavy work, such as excavation and blasting, along the entire line.[24]

The employment of soldiers for these tasks was ethically and economically flawed. Although they were not prison laborers, as their predecessors in earlier Soviet times often had been, they were definitely involuntary workers, ordered to their worksites by their commanding officers. They were assigned the most unpleasant jobs, the ones the voluntary workers did not wish to perform. Furthermore, they were paid very little, since they were drafted soldiers of the lowest ranks.

During most of the construction of BAM the use of soldiers as laborers was not discussed. After the advent of Gorbachev's *perestroika,* however (and before the completion of BAM), the Soviet press began to reveal the extent to which soldiers were being used. One writer castigated the practice on ethical grounds and also rightly noted that it distorted economic evaluations of construction:

The use of soldiers to do construction work in the civilian economy has long been a kind of "sacred cow" that could not be criticized in the press. But the fact is that using soldiers to fulfill plan assignments "corrupts" many of our departments, for their desires are no longer constrained by their resources—after all, their manpower costs them nothing.[25]

The employment of soldiers and short-sighted cost-cutting measures did not save the plan. The completion of BAM had been originally scheduled for 1983, but the actual completion did not come about until 2003. A particularly difficult section of the line was the Northern Muia Tunnel in Buriatia, at 9.5 miles the longest underground section on the route. Before the railway line had even been designed, geologists and engineers in Buriatia had warned the government that the intense seismic activity in the area made a tunnel inadvisable, and recommended instead a long bypass of the area.[26] Unwilling to accept the time delay the bypass's construction would cause, the government overruled the specialists and approved the tunnel.

Building the tunnel turned out to be much more challenging than expected. Pressured by their bosses to make 1984 the year to drive "the golden spike" symbolizing BAM's completion, the supervising engineers desperately threw together a 17.4-mile bypass that included grades too steep for regular freight trains to negotiate. A newspaper reporter who made the trip observed that "only a slalom skier could handle this road."[27] Nonetheless, the press could announce in 1984 that BAM had been completed. In truth, five more years would pass before the first regular freight trains could use the full length of BAM, and even then problems remained. Not until 2003 were most of these construction problems solved.

The construction of the great railway under such rushed conditions did considerable damage to the environment of Siberia.[28] The sewage produced by the hordes of construction workers had no place to go since in winter the creeks and small rivers often froze to the bottom and year-round the soil remained frozen a foot or two below the surface. Rivers and streams along the construction line were heavily polluted with oil, grease, rubbish, and discarded equipment. In winter the diesel engines of the heavy equipment were allowed to run all night, as otherwise they would not start in the morning. The engines created pervasive smog and pollution. Damage to the tundra of the region, which was exceedingly fragile, exposed the thin soil, which would not recover for decades. The indigenous Siberians, non-Slavic tribes, despised the Russians for killing the local game, ruining the land and water, and then leaving for home as soon as they had spent enough time on the site to earn their promised Zhiguli automobiles.[29] Lake Baikal, the oldest and deepest lake in the world, home to many unique species, became a major supply route for BAM; the lake abounded with transport boats in the summer and trucks on the ice in the winter. The newly awakened environmental movement vehemently protested the abuse of the lake.

BAM was operational by the early nineties, but its economic value remained dubious. The original hope that oil for export would be the most valuable freight on the railway was not realized because drilling for oil in that area of Siberia was so difficult it was indefinitely postponed. The development of the copper industry in Udokan, the second most important economic motivation for the railway, had in the meantime also been postponed. So far no large-scale mineral extraction from the area has been scheduled. The only significant cargo shipped the length of BAM has been timber, although it is more expensive than timber from elsewhere in Russia. Defenders of BAM often cite the profitable trade in coal from Southern Yakutia, though the coal does not travel on BAM itself but on a spur line connecting to the old Trans-Siberian Railway. A Russian economist observed in 1988, "For now there is nothing to haul on the new and expensive railroad, and BAM is an unprofitable venture."[30] It is quite possible that in the future, as that area of Siberia develops, the railway will become useful. Even if that happens, however, the manner of constructing the railway by fiat, disregarding a host of economic, environmental, and social consequences, will always be seen as a failure of rationality, at great cost.

The great dam on the Dnieper River, the enormous steel mill at Magnitogorsk, the White Sea Canal, the construction of industrial cities in the frigid far north, and the Baikal-Amur Mainline are all examples of the irrational, wasteful Soviet industrialization program that has bequeathed present-day Russia an industry that cannot begin to compete with the industries of other advanced countries. As a result, industry in Russia by early 2013 was wasting away, with many factories falling into ruin. Russia's economy today depends on trade (especially of oil, gas, minerals, and forest products), not industry. In a recent article titled "We do not manufacture anything," two Russian specialists maintained that factory production per capita in Russia today is ten times lower than in any other developed country.[31]

6 The Semiconductor Industry: Unheralded and Unrewarded Russian Pioneers

Transistors were one of the most important discoveries of the twentieth century, stimulating whole industries in much the same way that the steam engine did in the previous century. Almost no one in the West knows that a Russian inventor was the first person in the world to show that semiconductor crystals can amplify and generate high-frequency radio signals; the same man built solid-state radios in the 1920s and did important early work on LEDs, or light-emitting diodes. Decades later the few researchers in the West who learned of his work were astounded at the progress he had made toward transistors. Yet today, no Russian company ranks among the world's largest manufacturers of transistors, computer chips, or diodes. The reasons for this failure were political, economic, and institutional, not technical.

His intuitive choice and design of experiment were simply astonishing.
—American electrical engineer Egon Loebner, describing Oleg Losev's work on electroluminescence, thirty years ahead of his time

We are familiar with the very outstanding work done in solid state physics in the Soviet Union, and know the names of many of your scientists who have contributed so much to our knowledge.
—John Bardeen, Nobel laureate in physics, visiting Moscow in 1960

Semiconductors are at the heart of the revolution in electronic devices during the last sixty years. Transistors are types of semiconductors, and transistors are now used by the billions in communication, computer, and other devices, having long ago replaced most vacuum tubes of earlier years. They are excellent in regulating and amplifying electrical currents. Transistors have many advantages over tubes in most applications because of their small size, reliability, efficiency, and low cost. Semiconductor technology has boosted the intellectual power of humans much as the steam engine improved their physical capabilities. And just as the steam engine was probably the greatest invention of the eighteenth century, the transistor was probably the greatest invention of the twentieth century.[1]

Most people aware of the history of semiconductor technology date it to the post–World War II period. The invention of the transistor is usually credited to the Americans William Shockley, Walter Houser Brattain, and John Bardeen, whose work was announced by Bell Laboratories in 1948 (the three shared the Nobel Prize in Physics in 1956). Texas Instruments first placed a transistor radio on the market in 1954. But few people in the West are aware that a pioneer in semiconductor research was a Russian, Oleg V. Losev, who in 1922, a generation earlier, built working solid-state radio receivers and transmitters in Leningrad.[2] Although he had no formal university education, he conducted a full program of research that is documented in the scientific literature (a total of forty-three published papers and sixteen "author's certificates").

Losev was the first person in the world to show that semiconductor crystals can amplify and generate high-frequency radio signals.[3] In 1922 he built a radio using a zincite crystal and a cat's-whisker receiver made first of carbon and later of steel. It used very little power—just three or four flashlight batteries. The radio was known as the Crystodyne and was popular among radio amateurs, a burgeoning group in many countries. In the United States the monthly magazine *Radio News* published in 1924 an article stating:

Oscillating crystals are not new since they were investigated as far back as 1906 by well-known engineers, but it was not until lately that a Russian engineer, Mr. O. V. Losev, succeeded in finding some interesting uses for oscillating crystals. The construction of the apparatus by means of which oscillations may be produced with crystal as generator seems quite simple and should be of great interest to our readers.[4]

Losev's solid-state radio was a breakthrough, but it had drawbacks. Its range was limited and it was not very reliable, occasionally inexplicably ceasing to work. The theory of its operation was not understood. The Crystodyne was not at that time a true competitor to vacuum-tube radios, although radio amateurs loved to play with it.

Losev then went on to make another important discovery. Having learned how to generate radio signals with crystals, he began experimenting with different kinds of crystals in order to understand them better. He noticed in January 1923 that when he applied a current to silicon carbide, he "observed at the point of contact a weak greenish light."[5] At first he seemed not to attribute much attention to this phenomenon, working instead on radio receivers and transmitters. But then he began to study with a microscope the action of currents on silicon carbide. He experimented by changing polarities, varying the voltage, and probing different locations in crystals of a variety of composition and structure.[6] His ability to produce light in the crystal constantly improved, and he reported his results in the scientific literature in Russian, German, and English.[7] In Germany some scientists noticed his work and began speaking of "Lossev light." What Losev had done was to invent an LED, or Light-Emitting Diode. (The emission of such light was observed in 1907 by Henry Joseph Round but rediscovered by Losev, who went much further in exploring its characteristics.) Losev received an author's certificate for his "light relay" and believed it could be used for "fast telegraphic and telephone communication, transmission of images and other applications."[8] He employed Einstein's quantum theory to explain the action of the LED, calling it an "inverse photo-electric effect." He actually wrote to Einstein asking for help in developing the theory, but received no reply.[9]

Losev's life contains elements that are still not sufficiently explored. We know that his father was a former officer in the tsarist army and a nobleman.[10] With that kind of social background, Losev had to be very careful under the Soviet government, which was extremely suspicious of technical specialists it considered "alien" because of their social origins. Many such people ended up in prison. Despite his noble ancestry, Losev had no money and constantly looked for a source of income. For a brief time, the Soviet government gave him a chance. After an early period of economic militancy in which all private enterprise was eliminated the Soviet authorities relented a bit; during the New Economic Policy (NEP) period, from 1921 to 1927, some independent economic activity was allowed to flourish, especially small stores and enterprises. The government retained ownership of the "commanding heights" of the economy, especially heavy industry. These few NEP years were the ones during which Losev developed his Crystodyne solid-state radio, and therefore he saw a chance to sell them on the open market. In 1924 he advertised radio receivers and crystal detectors, and sold some.[11] We know that he made more than fifty of his radios.

However, a few years later a crackdown on private enterprise began, and Losev had to abandon all attempts to sell radios. He now had a doubly damning background: noble ancestry and practitioner of "bourgeois NEP-man economic activity." He tried to lie as low as possible. For a while he worked as a delivery boy for a radio institute, living in the stairwell that led to the attic. Yet he continued to do research, and his presence at the work-bench was tolerated by the other employees.

When the American (of Russian birth) electrical engineer Egon Loebner began working on electroluminescence in an RCA laboratory in the 1950s, he discovered some of Losev's scientific articles of thirty years before. He reacted in the following way: "His research was so exact and his publica-tions so clear that one has little difficulty determining today what he actu-ally did. . . . His intuitive choice and design of experiment were simply astonishing."[12] Loebner said that when he and his colleagues at RCA did research leading to commercially utilizable LEDs, they "followed the tech-nique of Losev."[13]

Losev developed a rudimentary theory of crystal amplification, speaking of "the penetration of free electrons in poorly-conducting layers" of a crys-tal. Crucially, he then made a silicon carbide crystal with four electrodes. He noticed that by applying a current to one pair, he could get amplification at the other pair. It is very tempting today to call this device a "transistor" (the word did not exist when Losev made these experiments). He presented his results at a scientific meeting in Leningrad and later published them in a journal called *The Herald of Electrotechnology*.[14] Fifty-six years later a cautious but admiring Russian physicist read this article and wrote, "Losev here was extremely close to the realization of the transistor because he had found a change in the conductivity between a pair of contacts when an electric cur-rent is applied to another pair."[15]

Should Losev be credited with inventing the transistor? That would not quite be justified, since Losev worked in an empirical way and had no knowledge of the physical theory behind transistors, although he had a remarkable intuition about what was happening on his laboratory bench. He could not fully explain what he had done—such explanations would come much later. What he needed to go forward was help from theoretical physicists and a secure place of research and development where his ideas could be fully tested and refined. But that was not to happen, and there are explanations for that lack.

Where would Losev find the supportive environment he needed? In the West, the two most likely possibilities were private industry or an academic laboratory, probably in a university. Both possibilities were closed to Losev. He could no longer sell radios, and by the 1930s private industry simply did not exist in the Soviet Union. The author's certificates (often called "patents") that Losev received for his inventions gave him no monopoly right in a commercial sense; Losev's innovations belonged to the Soviet government, which did nothing with them. Academic research establishments *did* exist in Soviet Russia at that time, and some were very good, but Losev was not qualified for them since he had never graduated from a university and had no advanced degrees. (A leading physicist, Abram Ioffe, later managed to get him an essentially honorary degree, but it came late, in 1938, and was not enough for Losev to acquire the kind of position he deserved.) Forced out of his position at the radio institute, he ended up as a teacher in a medical facility, where his interests were not appreciated or rewarded. He tried to continue his research, but between 1935 and 1940 he was not able to publish a single article. Although a talented man, he did not know how to promote his own interests (or perhaps hesitated to do so), and tended to look down on "applied research," even though that was what he was best at. He yearned for a position in a theoretical research institute, possibly seeking a sanctuary from the economic and political pressures that beset Soviet industry.[16] He was unsuccessful at social relations and had two failed marriages. When World War II came, this loner was offered a chance to leave the city before the Germans surrounded it, but he declined. He eventually died of starvation during the siege of Leningrad. He was thirty-nine years old, and might have worked for decades more under normal circumstances. But Losev was not a member of the scientific establishment in Russia, and he was not destined for normal circumstances.

The leader of the Russian scientific establishment in the field of semiconductors was Abram Ioffe, a member of the Soviet Academy of Sciences and director of the famous Physical-Technical Institute in Leningrad (now known as the Ioffe Institute in St. Petersburg). Ioffe's institute has often been called "the cradle of Soviet physics" and was the scientific birthplace of many distinguished scientists, including a number of Nobel laureates.

After the early 1930s, Ioffe directed a whole section of this institute devoted to semiconductors.[17] Later, after World War II, he established a separate institute of semiconductors. A typical Soviet policy was that whenever

Russian scientists noticed that a certain topic was becoming important in international science, they would establish a separate institute in that field. A dozen or more institutes were spun off from Ioffe's home institute in Leningrad. While these institutes often were places where good theoretical science was done, scientists working there almost never succeeded in creating commercial enterprises that prospered in the international marketplace. Ioffe was often praised by the regime (he won a Stalin Prize), but he faced an undercurrent of criticism during his entire life from knowledgeable observers, who said that very little good applied science ever came out of his laboratories, with the exception of some military applications, and almost no commercial technology of significance in the world market.

The undercurrent of criticism was correct. The history of semiconductor physics in the last two generations contains numerous moments when Russian scientists made important theoretical contributions. When John Bardeen, who, along with Shockley and Brittain, is usually credited with inventing the transistor, visited the Soviet Union in 1960, he observed, "We are familiar with the very outstanding work done in solid state physics in the Soviet Union, and know the names of many of your scientists who have contributed so much to our knowledge."[18] But a different story—and a more important one from the standpoint of the fate of nations—is that the history of industrial semiconductor technology assigns Russia very little role. Russia today is a giant in theoretical physics, including semiconductor physics, but it is a midget in industrial high technology on the international level. No more dramatic illustration of the gap between Russian scientific achievement and industrial technology can be found than in the field of semiconductors.

In the early twenty-first century the director of the Ioffe Institute was Zhores I. Alferov, winner of the Nobel Prize in Physics in 2000 for the development of semiconductor heterostructures. Despite Alferov's achievement, Russia continued to be very backward in the commercial exploitation of transistors. Recognizing this deficiency, the Russian government in 2010 decided to create its own version of Silicon Valley, called Skolkovo. They appointed as Russian chair of the scientific advisory committee of Skolkovo the director of the Ioffe Institute, Alferov, a member of the Communist Party and a defender of the traditional organization of Russian science, with academy institutes the centerpieces. The Russian government could have made a much better choice for the head of the Skolkovo advisory council than Alferov, who was conservative both in politics and in scientific viewpoint. An illustration of Alferov's political views came in June 2012, when he praised the development of the republic of Belarus and commented that "Belarus is a modern, civilized European country." This is the same country that former U.S. secretary of state Condoleezza Rice called "the last true remaining dictatorship in the heart of Europe."[19]

Today no Russian company ranks among the world's largest computer or computer chip manufacturers. The worldwide electronics industry based on the transistor is one in which Russia plays a surprisingly small role.

In contrast, in the United States the connections between transistor research, academia, and industry were intimate. The invention of the transistor was announced by Bell Labs, a part of AT&T, where both William Shockley and Walter Brattain worked. Several of the American pioneers in transistor research became heavily involved in industry. Shockley started his own company, and although it failed, several of the people who worked for the company broke off and founded other firms, including Intel, a giant in the field today. Bardeen served as a consultant to several industries and for many years was on the board of Xerox. These three men, who shared the Nobel Prize in Physics in 1956, later went their separate ways. Shockley irritated Brattain and Bardeen with his claims to priority, but all three remained committed to the union of academic research and private industry in very practical ways, something missing in the Soviet Union.

7 Genetics and Biotechnology: The Missed Revolution

A brilliant school of biologists and geneticists developed in early Soviet Russia. These scientists first presented the concept of a "gene pool" (a Russian term) and made a significant contribution to the "modern synthesis" bringing Mendelian genetics and Darwinian evolution together, a necessary step for the further development of modern biology. A Russian botanist in these years was the first scientist to actually create a new species. These Russians worked closely with the leading biologists and geneticists of other countries. One of these foreign biologists, the American future Nobel Prize–winning H. J. Muller, was so impressed with the Russian work that he learned the Russian language and traveled to the Soviet Union to work with his Russian colleagues. However, this outstanding school of biology was wiped out by a political campaign in the Soviet Union headed by Trofim Lysenko, who did not accept modern genetics. Backed by the power of the government and police, Lysenko suppressed modern genetics in the Soviet Union. Although biological teaching recovered in the post-Soviet years, even today the results of this catastrophe are observable. At present, Russia does not have a single biotechnology company in the top one hundred in the world in terms of revenue.

It continues to be very interesting here and I can see that there are great possibilities for the development of research in genetics.

—Hermann J. Muller, American future Nobelist in biology, writing from Leningrad in 1933[1]

In the 1920s a talented group of biologists in Russia was laying the foundations for what could have been a brilliant school of genetics leading to molecular biology, a field that has transformed biological science and biotechnology worldwide. After an outstanding start, Russia played almost no role in this revolution, and although currently it is making great efforts to catch up, it still lags far behind. The reason for this momentous failure after great early promise was almost entirely political. For a few brief years, however, Soviet Russia was at the forefront of genetics research. Russian biologists helped create the evolutionary synthesis that underlay the further development of the field.[2] And in 1927 one of these Russian biologists, Georgi Karpechenko, created a new plant species.[3]

To understand this achievement of Soviet biology and the subsequent slump, a bit of the history of biological studies is helpful. In the early twentieth century many biologists saw a tension, if not a contradiction, between Darwinian evolution and Mendelian genetics. Darwinians envisioned changes in organisms occurring gradually over long periods of time, based on minute variations; Mendelists at first emphasized the extreme stability of the gene, then later, incorporating the concept of mutation, depicted stability occasionally interrupted by rather large changes, a different picture from the traditional Darwinian one. Furthermore, the members of the different camps proceeded on the basis of contrasting methods. The followers of traditional Darwinism emphasized descriptive natural history, while the new Mendelians used mathematical approaches. Was there any way these two intellectual strands could be brought together, or was the Darwinian approach destined to be superseded by the Mendelian?

By the second decade of the twentieth century biologists concerned with these problems could be roughly classified into three groups: naturalists working within the late nineteenth-century Darwinian tradition; geneticists studying gene location and mutation, many of them connected with the school of T. H. Morgan at Columbia University in New York; and "biometricians," who used highly mathematical methods as developed by Karl Pearson and others. While some hope existed of finding a commonality or synthesis of the different approaches, the means was not apparent.

One of the most important papers pointing the way to such a synthesis was written by a Russian, Sergei Chetverikov, in 1926.[4] Chetverikov noted in the opening sentences of his article that Mendelism was greeted with hostility by "outstanding evolutionists" both in Russia and abroad, then stated that his goal was to bring these two approaches together through a clarification of evolution from the perspective of genetic concepts. Chetverikov went on to argue that the process of mutation being observed in laboratories also occurred in nature, but that because recessive mutants would be heterozygous, they would not be evident in phenotypical forms. Natural selection would quickly eliminate harmful dominants but would act more slowly against harmful recessives. Thus, there would be a build-up of hidden recessive mutants in any population.

In the same paper, Chetverikov agreed with the American biologist T. H. Morgan and others that selection cannot affect genes themselves, but he emphasized that genes do not act in isolation from the rest of the genotype:

The very same gene will manifest itself differently, depending on the complex of the other genes in which it finds itself. For it, this complex, this genotype, will be the genotypic milieu, within the surroundings of which it will be externally manifested. And as phenotypically every character depends for its expression on the surrounding external environment, and is the reaction of the organism to the given external influences, so genotypically each character depends for its expression on the structure of the whole genotype, and is a reaction to definite internal influences.

Chetverikov was presenting here extremely sophisticated concepts of population genetics, and he and his students supported these concepts with original experimental work on natural populations.

Another Soviet contribution to genetics in these years was the concept of gene pool. The Russian geneticist A. S. Serebrovskii first formulated the concept in terms of *genofond* (gene fund), a word that was imported to the United States from the Soviet Union by Theodosius Dobzhansky, one of the members of Chetverikov's group, who translated it into English as "gene pool." Today few people know that this term, so common in biological discourse the world over, has a Russian origin. Yet another Soviet researcher, a student of Chetverikov's, D. D. Romashov, arrived independently at the concept of genetic drift, developed in the West by Sewall Wright and others. And still another, Yuri Filipchenko, coined the terms "microevolution" and "macroevolution" and brilliantly incorporated Mendel's laws into evolutionary theory, thus giving the modern synthesis a great boost.

Chetverikov, Serebrovskii, Filipchenko, Dobzhansky, Karpechenko, and Romashov were among the many pioneering Russian biologists of the 1920s.[5] Others included Nikolai Kol'tsov, N. V. Timofeev-Resovskii, Nikolai Vavilov, and N. P. Dubinin. They were working closely with leading biologists and geneticists in other countries, such as Hermann J. Muller from Morgan's lab at Columbia University, a man who would eventually win the Nobel Prize in Physiology or Medicine for demonstrating X-ray mutagenesis. Muller was so impressed with the work of this group of Soviet geneticists that he visited and worked with them frequently, at one point for several years.

If these Soviet geneticists had been able to maintain their pioneering school, producing graduate students for the next generation, Russian biology would doubtlessly have flourished and Russia might have become a full participant in the molecular biology revolution that swept the rest of the scientific world and led to great practical results, creating whole industries. But this was not to be.

Almost all of the Russian biologists of the twenties suffered from Stalinist repression. Chetverikov was arrested and sent into exile and never returned to his main research topic. Dobzhansky emigrated to the United States to escape political controls and became a famous scientist at Rockefeller Institute (later Rockefeller University). Karpechenko was sentenced to death and executed in 1941. Kol'tsov was accused of ideological sins and dismissed from his position, and left the field. Vavilov was arrested in 1940 and died of malnutrition in prison in 1943. Dubinin abandoned genetics in 1948 and worked for many years as an ornithologist, returning to his main studies only after 1965. Romashov was arrested twice but released because of illness; his wife died in prison. Timofeev-Resovskii emigrated to Germany, returning to Russia only many years later.

The reason for all these difficulties was political repression and the rise of a fraudulent view of genetics promoted by a poorly educated agronomist, Trofim Lysenko. Lysenko effectively destroyed modern genetics in the Soviet Union and caused untold damage to Soviet agriculture. This book is not the place for the history of how that happened (the story is well told elsewhere,[6] and I have also published the story of my meeting and talking with Lysenko in 1971[7]), but one correction to the most common interpretation of these events should be made: contrary to much speculation in the West, Lysenko never preached, on the basis of the inheritance of acquired characteristics, that he could "create a new Soviet man." In fact, Lysenko never claimed that his views on heredity were applicable to human beings. He derived his strengths from his false claims that he could improve agriculture, a crisis area of the Soviet economy, and, later, from his attempt to show that his biological views accorded with Marxism. What were his biological views? He believed that through a process he called "vernalization" he could vastly improve the production of wheat and potatoes and, later, corn and dairy products. He even claimed that he could convert winter wheat into spring wheat.

By appealing directly to Soviet leaders such as Stalin and Khrushchev, who were very eager for agricultural improvement and incapable of judging the validity of such claims themselves, Lysenko won political support for his agricultural nostrums. These political leaders not only supported Lysenko but suppressed his critics, biologists who knew that Lysenko was a fraud. Several thousand of these biologists were eventually arrested, and a few, such as a leading biologist at Moscow University, D. A. Sabinin, committed suicide.

The story of genetics in Soviet Russia is obviously a human tragedy, but in terms of "Russia's unfulfilled promise" that this book emphasizes, it is also another example of the familiar story of achievement followed by failure in many promising technologies. Instead of becoming a world leader in scientific agriculture and biotechnology on the basis of its early successes in genetics in the twenties and thirties, Russia became, once again, a follower of progress elsewhere. No Russian company today is a major international player in the field. In fact, Russia today does not have a single biotechnology company in the top one hundred in the world in terms of revenues, although companies based in twelve different nations are on that list.[8]

8 Computers: Victory and Failure

Russians were pioneers in the early development of calculators, computers, and the mathematical foundations of information science. In the late tsarist period Russian engineers and scientists made important advances in calculating devices. In the Soviet period a whole group of mathematicians—V. A. Kotelnikov, Andrei Kolmogorov, I. M. Gelfand, and others—made significant contributions to information theory. Furthermore, Soviet scientists and engineers built the first electronic digital computer in continental Europe. When American and Soviet engineers first began cooperating on the space program, Soviet engineers were sometimes able to solve computer problems more rapidly than their American colleagues. However, in subsequent years computers increasingly became commercial objects, and the Soviet Union was not able to keep up in the market competition. Soviet computer scientists were forced to abandon their independent efforts and adopt IBM standards. Today there is no significant Russian computer manufacturer in international competition.

He used to take his papers and a candle to the bathroom, where he was writing "ones" and "zeroes" for hours.
—Alisa Grigorevna Lebedeva, wife of Sergei Lebedev, Soviet computer pioneer, describing his behavior in Moscow in 1941 during German bombing raids

Russians were active very early in the development of calculating devices, information theory, and computers. Even before the Russian Revolution of 1917, Russian engineers and scientists made important advances in this field. The Russian naval engineer and mathematician Alexei Krylov (1863–1945) was interested in applying mathematics to problems of shipbuilding. In 1904 he built a machine to solve differential equations. Another younger engineer, Mikhail Bonch-Bruevich (1888–1940), working in the same city,

St. Petersburg, was studying vacuum tube development for radios. Around 1916 he invented one of the very first flip-flop relays based on an electronic circuit with two cathode ray tubes.

In the West, one of the pioneers in information theory was Claude Shannon, who in 1937 wrote a master's thesis at MIT in which he showed that it would be possible to use arrangements of relays and binary arithmetic to solve Boolean algebra problems. Shannon's work was the foundation for digital circuit design for computers. Few people in the West know that two years earlier, in 1935, the Russian logician Viktor Shestakov had proposed a similar theory of electric switches based on Boolean algebra, but published his work only in 1941, three years after the publication of Shannon's thesis. In this early period neither Shannon nor Shestakov was aware of the other's work.

The first electronic computer in continental Europe was built in secret in 1948–1951 in a ramshackle group of buildings in a place called Feofania, near Kiev. Before Soviet times the buildings had been a monastery, surrounded by oak woods, carpets of flowers, wild berries, and mushrooms, and numerous wild animals. During the early Soviet years the monastery was converted into a psychiatric hospital (a frequent Soviet practice was to convert religious institutions into research or medical facilities). All the patients in the hospital were killed or dispersed in World War II, when the buildings were badly damaged. The muddy road to the location was nearly impassable in the spring and fall and was a bumpy adventure even in the best of seasons. In 1948 these half-ruined buildings were turned over to Sergei Lebedev, an electrical engineer, for the building of an electronic computer.[1] In the Feofania research facility Lebedev and about twenty engineers, along with ten assistants. built the MESM (the Russian initialism for "small electronic calculating machine"), a computer that was one of the fastest in the world and exhibited many interesting features. Its architecture was entirely independent of that of the only better computers in the world at that time, the few being developed in the United States.

Sergei Lebedev was born in the city of Nizhny Novgorod (later Gorky, today again Nizhny Novgorod) in 1902. His father was a schoolteacher who was transferred from place to place, and as a result, Sergei lived in a number of towns while growing up, mostly in the Urals region. Eventually his father was transferred to Moscow, and there Sergei entered one of Russia's biggest engineering schools, the Moscow Higher Technical School, known today as the Baumann Institute, where he became interested in high-voltage engineering, a field requiring mathematical preparation and ability. Upon graduation he became a lecturer at the Baumann Institute and a researcher in the Laboratory of Electric Networks. He was an avid mountain climber and later named one of his computers after the highest mountain in Europe, Elbrus, in the Caucasus, which he successfully climbed.

In the late 1930s Lebedev became interested in binary arithmetic. During the autumn of 1941, when Moscow was blacked out because of German bombing raids, his wife, a musician, recalled that he "used to take his papers and a candle to the bathroom, where he was writing 'ones' and 'zeroes' for hours." Later in the war he was transferred to Sverdlovsk (now Ekaterinburg), where he worked on the design of munitions. He found that he needed a calculating device that would solve differential and integral equations, and in 1945 he created Russia's first electronic analog computer, but he already had the idea of a digital computer based on binary arithmetic. Remarkably, so far as we know he was not familiar at this time with the works in this field of either his countryman Shestakov or the American Claude Shannon.

In 1946 Lebedev was transferred from Moscow to Kiev, where he began work on an electronic computer. In 1949 a leading mathematician and science administrator in Kiev who was familiar with Lebedev's work, Mikhail Lavrentiev, wrote Stalin urging the acceleration of the development of computers, emphasizing their utility for military purposes. Stalin replied by giving Lavrentiev authority to establish a new computer science laboratory, and Lavrentiev invited Lebedev to head it. Now Lebedev had the advantages of money and priority. At the same time, this order by Stalin illustrated the importance of political authority—actually, the importance of one man—in the promotion of technology in the Soviet Union.

Lebedev developed the MESM only three or four years after the world's first electronic computer, the ENIAC in the United States, and at about the same time as the British EDSAC. By the early 1950s the MESM was being used to solve problems in nuclear science, space flight and rocket engineering, and remote power supply transmission lines.[2]

In 1952 Lebedev followed the development of MESM with another computer, the BESM (the initialism for "big electronic calculating machine"). It was the fastest computer in Europe and, at least for a while, comparable to the best computers in the world, a genuine triumph. Only one BESM-1 was built, but later models, especially the BESM-6, were produced in hundreds of units and were used in many applications. Manufacture of the BESM-6 computer ended in 1987. In 1975, during the Apollo-Soyuz space cooperative project between the United States and the USSR, the processing of Soyuz orbit parameters was accomplished by a BESM-6 based system more rapidly than the Americans could do the same calculation.

And yet, after this promising start, Russian computers fell far behind the best in other countries. This failure can only be understood by reviewing the history of the computer industry, taking full account of the social and economic factors that eventually transformed it. The computer industry after World War II was shaped in leading nations by three major forces: academia, the government (especially military concerns), and business. The forces of academia and the government were especially important in the earlier years, with business playing a fundamental role only later. The Soviet Union did just fine in computers so long as the development of these devices was determined largely by academia and the government. The Soviet government was unstinting in its support of computers for air defense and research on nuclear weapons. But then business became a major force in the West, marked symbolically by General Electric's decision in 1955 to buy an IBM 702 computer to automate its payroll and other records in its Schenectady plant and Bank of America's decision in 1959 to computerize its banking industry (using an ERMA computer designed by the Stanford Research Institute).

These efforts inaugurated the widespread computerization of banking and business. In the 1960s and 1970s computers became commercial commodities, with all the reductions in costs and improvements in ease of use that market forces produce. The Soviet Union, with its centralized, noncompetitive market, could not keep pace with the advances being made, and in the 1970s the Soviet computer industry gave up its earlier and impressive effort to pursue an independent path and instead adopted IBM standards. From that point on, Russians were followers in computers, not leaders.[3] Sergei Lebedev died in 1974, but another leading designer of Soviet computers, Bashir Rameyev, lamented the decision to adopt IBM architecture until his death in 1994.[4] What defeated the Soviet computer industry was not lack of ability or knowledge in computer science, it was the overwhelming force of the marketplace.

Another factor, although not a determining one, was ideology.[5] In the 1950s Soviet ideologists were very critical of cybernetics, calling it "the science of obscurantists." In 1952 a Marxist philosopher condemned the field as a "pseudoscience"and ridiculed the belief that computers could help explain human thought or help understand social activity.[6] In another article published the following year, titled "Whom Does Cybernetics Serve?," an anonymous author ("Materialist") condemned cybernetics as contradicting Marxist dialectical materialism and described the science of computers as a particularly pernicious effort by Western capitalists to extract more profits from industry by eliminating the need to pay wages to the proletariat.[7]

Although such ideological pronouncements may have had some depressing effects on computer science in the Soviet Union (such as discouraging young people from entering the field), the development of computers, especially for military uses, continued unabated.[8] As one Soviet computer scientist told me in 1960, "We did cybernetics without calling it cybernetics." Furthermore, in the late 1950s and early 1960s, the Soviet Union did a U-turn on cybernetics and began celebrating it as a science that could serve the goals of the Soviet Union. An important 1961 book was even titled *Cybernetics in the Service of Communism*.[9] Many Russian universities and research institutes developed departments of cybernetics, and some still have them today.[10]

A more serious political threat to the development of computers in the Soviet Union came with the emergence of personal computers.[11] The Soviet authorities liked computers so long as they were big mainframe machines in central governmental, military, and industrial offices but were less enthusiastic when computers moved into private apartments, where they could be used for the rampant dissemination of information by ordinary citizens. In an effort to control information, the Soviet authorities had long banned the private possession of printing presses and copying machines. A personal computer with a printer was the equivalent of a small printing press. What should Soviet authorities do about this?

The peak of the debates among Soviet leaders about computers came in the mid- to late 1980s. In 1986 I talked about this problem with one of the top computer scientists in the Soviet Union, Andrei Ershov. Although far from being a dissident, he was surprisingly frank, agreeing that the desire of the Communist Party to keep control over information was hindering the development of the computer industry. He then made the following statement:

Our leaders have not decided whether a computer is most like a printing press, a typewriter, or a telephone, and much will depend on their decision. If they decide that computers are like printing presses they will wish to continue controls, just as they do over all printing presses at the present time. Individuals will not be able to own them, only institutions. If, on the other hand, our leaders decide that computers are like typewriters, individuals will be able to own them and the authorities will not try to control the actual machines, although they may try to control the distribution of the information produced by them. If, lastly, our leaders decide that computers are like telephones, most individuals will have them, and they will be able to do with computers what they wish, but their on-line transmissions will be occasionally monitored.

I am convinced that eventually the Soviet government will have to permit personal computers to be owned and controlled by individuals. Furthermore, it will become obvious that personal computers are not like any previous communication technologies—not like printing presses, nor typewriters, nor telephones. Instead, they are a totally new type of technology. The time is coming soon when any individual anywhere in the world will be able to communicate almost instantaneously with any other individual anywhere in the world. That will be a revolution—not only for the Soviet Union, but for you too. But its effects will be greatest here.[12]

This statement illustrates graphically how difficult a political problem computers were for the Soviet leaders. But the question quickly became moot because five years after Ershov made this statement to me the Soviet Union disappeared, and all controls over communication technologies were abandoned (but not controls over mass media, such as television).

In post-Soviet Russia the computer industry has never recovered from the retardation it suffered in the late Soviet period. As we have seen, that retardation was caused more by inability to compete with market forces than it was by political controls, but the latter also played a role. Today there is no Russian computer manufacturer that is significant at the international level, even though Russians can justifiably claim to have been among the pioneers in the field of computer science.

9   Lasers: Genius and Missed Opportunities

Russians were pioneers in the development of lasers, today a multi-billion-dollar industry. Two of them, Alexander Prokhorov and Nikolai Basov, won the Nobel Prize in 1964, along with the American Charles Townes, for the invention of lasers and masers. Even earlier, in the 1930s and 1940s, the Russian scientist Valentin Fabrikant laid the foundations of physical optics and gas discharges that led to the development of lasers.

Curiously, the history of lasers illustrates the strengths and weaknesses of both the American and Soviet systems, since political and economic obstacles were present in both systems and slowed development in both countries. However, investor interest and commercial competition were much stronger in the United States than in the Soviet Union, leading to the development of important American laser start-up companies. Today no Russian manufacturer of lasers is a major player internationally.

They did not do their homework. It would have made more sense to recognize the Russian physicist Fabrikant.

—Theodore Maiman, the American who built the first laser, speaking of the awarding of the 1964 Nobel Prize for the development of lasers to Charles Townes, Alexander Prokhorov, and Nikolai Basov.[1]

In April 1955 a physics professor at Columbia University in New York City named Charles Townes journeyed to a scientific meeting in Cambridge, England, where he wanted to talk about the research he had been doing on stimulated microwave radiation. About a year earlier, Townes, working together with several graduate students in Pupin Hall at Columbia University, had successfully demonstrated such radiation in a device in which ammonia molecules were bombarded with microwaves. The result was an output of only several billionths of a watt, but it showed that the device worked. Townes had made a breakthrough.

Together with his students over lunch after the momentous event, Townes struggled to come up with a name for the device, and settled on "maser," or "microwave amplification by the stimulated emission of radiation." Townes's ammonia maser was the ancestor of a quickly following generation of devices, including, a few years later, the more ambitious "laser," or "light amplification by stimulated emission of radiation." Lasers and masers would become a multi-billion-dollar industry and would be at the heart of a multitude of modern electronic gadgets that practically everyone uses today.[2]

After his achievement, Townes sent a short paper announcing it to the leading journal *Physical Review*, but at the time of his trip to England he had not yet written a thorough theoretical description of what he had done. Therefore, he was astounded when a Soviet physicist named Alexander Prokhorov,[3] speaking excellent English, preceded him at the conference by delivering a paper describing the theory of an ammonia maser, exactly the device Townes had used. Townes had never met Prokhorov before and did not dream at that moment that he would eventually share a Nobel Prize with him and his student, Nikolai Basov,[4] for the development of lasers. Townes's first concern in Cambridge was to assert his claim to his invention and not yield it to Prokhorov. After Prokhorov had finished his report, Townes stood up and announced, "Well, that is very interesting, and we have one of these working." He then described his recent work with an ammonia maser.

When people began to study more closely the work that Russian scientists had done in the field of physical optics and gas discharges, they were amazed by what they found. Theodore Maiman, the American who built the first laser in 1960, later observed, "It was the Russian physicist A. V. Fabricant [sic] who first had the vision to propose the concept of a laser in 1940."[5] Fabrikant stated the principles of a laser in a doctoral dissertation of 1939, and in 1951 he obtained for his work an "author's certificate" (often called a "patent" in the West, but something very different from a Western patent). He not only elaborated the theory, he was the first to observe experimentally the amplification of optical radiation using a mixture of mercury vapor and hydrogen. He was a true pioneer.[6] Maiman even thought Fabrikant should have received the Nobel Prize instead of Basov,

Prokhorov, and Townes. He said that when the Nobel Committee chose the trio over Fabrikant, they "did not do their homework. It would have made more sense to recognize the Russian physicist Fabricant [sic]."[7]

Authors' certificates in the Soviet Union carried no monopoly or financial rights for Russian inventors; they were merely honorific recognitions, although sometimes they came with a modest one-time financial bonus. However, they gave inventors no ability to market and financially benefit from their creations. And in fact, such benefit could not have been further from Fabrikant's mind. He admitted that he "did not pay attention to the practical value of his idea."[8] He was a typical Russian *intelligent*, an erudite conversationalist, a person who lived in a world of ideas. Business was simply not his concern.

The development of the laser in the United States and the Soviet Union displayed the strengths and weaknesses of the two systems and the close connections of both to military interests. In the United States, private inventors and commercial companies such as Hughes Aircraft and AT&T were promoting laser research and hoping for patents and commercial benefits. Later, nasty patent disputes dragged on for decades among these people and organizations.[9] Since the laser was the object of attention of military circles, where it was thought to hold promise of becoming a powerful weapon, questions of secrecy and security clearance soon arose, with effects that actually delayed research. In the Soviet Union the centralized research system favored high-priority projects such as laser research, and the Soviet military quickly became involved, supplying necessary funds. When in later years I interviewed Prokhorov in his Moscow lab, he proudly announced that not only had he invented the laser, he had also discovered a "general effect" in physics. When I asked what the "general effect" was, he replied, "Military generals are very interested in my physics."[10]

As the recipient of a Nobel Prize in Physics for work on masers and lasers Charles Townes is today most often cited as the inventor of the laser, but priority in this achievement is still hotly disputed by physicists, commercial companies, and historians of science and technology.[11] Other candidates for the accolades, along with Townes, were his brother-in-law Arthur Schawlow, Theodore Maiman of Hughes Research Laboratories, the ambitious inventor Gordon Gould, four different groups of physicists working at Bell Laboratories, and several groups in the Soviet Union. These scientists sniped at each other for decades over priority.

On May 16, 1960, Theodore Maiman, working at Hughes Research Laboratories in Malibu, California, first operated a laser, one based on a synthetic ruby crystal.[12] Soviet researchers were very close behind.[13] Two groups, one at the State Optical Institute in Leningrad and the other at the Lebedev Physics Institute in Moscow, were already experienced in the field, and as soon as they heard of Maiman's achievement they attempted to duplicate it. The Leningrad group, headed by D. D. Khazov, succeeded on June 2, 1961, and the Moscow team, composed of M. D. Galanin, A M. Leontovich, and Z. A. Chizhikova, did the same thing a few weeks later, on September 15.

The big difference between Maiman and his Soviet competitors is what they did *after* building the lasers. Maiman's company, Hughes Aircraft, took out a patent, which later became very lucrative. Maiman, deprived of financial benefit since as an employee of Hughes he was required to give the company patent rights, became disgruntled, left the company, and became president of his own laser company. None of the Soviet researchers did anything similar, nor could they, since the Soviet Union had neither patents nor private companies. More important, it never even occurred to the Soviet researchers to try to commercialize their work.

The history of the early development of lasers in the United States and the Soviet Union is very revealing, demonstrating clearly the strengths and weaknesses of the two systems. The Soviet system stymied innovation and commercial development, but the American system posed barriers to innovation as well. Particularly interesting is the case of Gordon Gould, an American researcher who ran head-on into the political problems of the United States stemming from its competition with the Soviet Union.

Gould, born in 1920, earned a master's degree in physics at Yale University. He was involved in the Manhattan Project but was dismissed from it because of his association with left-wing organizations. He went through a radical period and, along with his even more radical first wife, joined Communist organizations. After about 1950, however, he became disillusioned with the Soviet Union. He went to Columbia University in 1949 to work for a PhD in physics under Polykarp Kusch, a close associate of Charles Townes and an eventual Nobel Prize winner himself.

Gould differed from Kusch and Townes, however, in that he envisioned himself more as an inventor than as an academic physicist (he never finished his PhD, and his hero was Thomas Edison, not his professors). Despite his left-wing inclinations, he wanted to get rich from his inventions. He became very interested in masers and optics, and often talked to Townes about his research. It was actually Gould who first coined the term "laser"; Townes called the device he was working on an "optical maser."

In 1957 Gould secretly compiled a notebook combining the research emerging in the Columbia lab with his own fertile ideas, and then had it notarized (each page separately, illustrating how important he thought the notebook was) at a local store where the proprietor was a notary public who understood nothing about the contents. That notebook would be a basic document in the later patent disputes.

The next year, 1958, Gould left graduate school and joined a small private company, TRG (Technical Research Group), and pushed his laser project. In 1959 the company was successful in getting a large grant for research on lasers from ARPA (Advanced Research Projects Agency, a U.S. government organization pushing projects with military potential). On the one hand, this development was a great achievement for Gould; on the other hand, it became the source of problems he confronted the rest of his life. ARPA insisted on classifying the research, but Gould, because of his radical background, could not get security clearance. Soon the ridiculous situation developed in which the person who was considered the leader of company research in lasers, Gordon Gould, was not permitted in the laboratories where the research, now classified, was being done. In fact, Gould's office was placed across the street from the lab, and he was not permitted to cross the boundary.

Although no one can predict what would have happened if this absurd situation had not arisen, most people agree that it slowed research progress. Gould, who paid little attention to rules, matched the deadly foolishness imposed by the government on his lab by having an affair with the woman officer in the company who was in charge of security and top-secret materials. When that compromising fact was learned, the woman was fired; Gould was not. On another occasion Gould stole a laser from the classified section after hours. But he was still valued for his skill as an inventor (he worked on many projects other than lasers, including contact lenses and dental devices).

Gould never received security clearance, although he tried for years. Eventually he became an idiosyncratic rebel, convinced the establishment was against him, and he adopted the strategy of multiple patent suits, trying to get royalties from the booming laser industry. Townes, a proper southern gentleman, and others believed that Gould was trying to take credit for work done by others. Gould in return accused Townes of using his ideas.

At first Gould was unsuccessful in his patent litigations, but eventually he won a crucial suit and ended up making millions of dollars from the laser industry. The bizarre result of Gould's adventures is that the man who was a radical with communist sympathies in his youth ended up in his older years as the most successful capitalist of the early laser researchers. Other researchers resented his success, which they considered unjustified; Maiman called Gould's winning patent suit "a travesty of justice."[14] He had little better to say of Townes, who was, in his opinion, a part of the "old boy establishment," which would not make room for an industrial scientist who had not come from a prestigious Eastern university.[15]

Alexander Prokhorov, leader of Soviet research in the field, was also a man of very strong opinions, and he did not shrink from imposing them on others. Many stories have been told about his authoritarianism, and one of them was so dramatic that, although I had heard it many times, I had trouble believing it. Then in 2007 the minister of science and education of Russia, Andrei Fursenko, repeated the story in a major Russian newspaper.[16] The story was this: In the 1950s Prokhorov's lab at the Lebedev Physics Institute was doing rather conventional research that did not seem to be leading toward anything exciting. Prokhorov decided it was necessary to move in a different direction, to start working on induced radiation of gases. His assistants in the lab did not wish to do so, as they were working on their dissertations and did not want disruption. Prokhorov gave them a month to rethink their positions. When they refused, he took the radical action of going through the laboratory with a hammer and destroying the instruments that were necessary for their current research. He then brought in new instruments and instructed the assistants to work on what he told them. A tremendous scandal ensued and half the researchers left, but the remaining ones followed Prokhorov in the work that eventually resulted in a Nobel Prize.

This personality trait could have negative as well as positive results. Prokhorov was a communist, belonging to the party since 1950, and a proud Soviet patriot. He could not countenance political disloyalty. Although he knew the Soviet dissident scientist Andrei Sakharov well, he strongly disagreed with his opinions. When Sakharov published an article criticizing

Soviet foreign policy, Prokhorov, along with three other colleagues, wrote a letter to the major Soviet newspaper *Izvestiia* castigating Sakharov in the strongest possible terms, calling him a traitor without honor.[17] This scurrilous attack badly damaged Prokhorov's reputation in the West. When Prokhorov came to San Francisco to give a paper, the American scientist Andrew Sessler marched in front of the podium with a sign which read "Prokhorov—great scientist, lousy human being."[18]

When one considers what an important part Russian scientists played in the development of the laser, a striking fact is how unimportant Russia is in the worldwide laser industry. By the year 2000, approximately $200 billion worth of lasers and laser systems had been sold.[19] Yet the Russian share of the world laser market at this time, thirty-six years after two Russians and an American were awarded the Nobel Prize for the invention of the maser and laser, was merely 1–1.5 percent.[20] The largest laser manufacturers at that time were American. No Russian manufacturer was a major player.

To explain this dramatic contrast between early Russian leadership in lasers and its later weakness in commercialization, it helps to consider where the world laser industry came from. Although some large companies manufacture lasers, most laser groups began as independent start-ups at a time when such small independent enterprises were impossible in the Soviet Union. The largest laser manufacturers in the early period were Spectra-Physics, Inc., and Coherent Inc., both established in the United States in 1961 and 1966, respectively. In 2004 Spectra-Physics was acquired by the Newport Corporation, another American company, founded in 1969.

Although the backstory of large companies being established in a garage in Silicon Valley is now almost a cliché, several of the largest laser manufacturers began exactly that way. The Newport Corporation was founded in 1969 by graduates of the California Institute of Technology, John Matthews and Dennis Terry, who were soon joined by another graduate of CalTech, Milton Chang. In their first year of operation they generated sales of $46,000—not very much, but enough to move out of the garage into leased industrial space. The little company soon acquired a high-tech customer base and manufactured precision optic, electro-optic, and opto-mechanical products and pursued interferometry and holography. The company went public in 1978.

The origin of another major laser manufacturer, Coherent Inc., illustrates that even the rebellious young founders of the laser industry were rebelled against. A young physicist named James Hobart was an early employee of the pioneering Silicon Valley laser company Spectra-Physics. He became interested in industrial lasers for use in such tasks as cutting and welding metal, but he could not convince his bosses to pursue these products.

In 1966, Hobart, in his early thirties, left his job and started his own company in Palo Alto, California, with start-up capital of $10,000. The new company was at first housed in a laundry room, where Hobart found a 220-volt outlet, necessary to power his new laser. Next to a washer and dryer, Hobart and his colleagues built their first industrial laser, using a rain gutter as one of its key components. That instrument was the first carbon dioxide laser available commercially. It was a success, and Hobart very quickly replaced the rain gutter with a shiny metal tube. Hobart managed to sell the laser to the Boeing Company, although he accidentally burned a hole in a sport coat during the successful demonstration. The company took off quickly and in 1970 went public. By the 1980s it was the world's largest independent laser maker.

No company in the laser business, no matter how gifted and independent its founders, was safe from creative dissent and possible division. One of the founders of the Newport Corporation, John Matthews, in the 1970s developed a laser sight for firearms. Unable to find the funds he needed for further development within his company, he resigned and started a new company, which became Laser Products, later SureFire.

Thus we see that the early laser industry in the United States was promoted by rebellious spirits. Some succeeded, many failed. In the late 1980s, after numerous successes, the Coherent Corporation almost went bankrupt, besieged on all sides by competitors. The laser industry was a dramatic illustration of "creative destruction," a term that, ironically, first came from Marxist economic theory but now is more often associated with the economist Joseph Schumpeter.

Because of its centralized economy the Soviet Union could not develop laser companies in the individualistic, competitive, and, yes, chaotic way in which they arose in the United States. However, the Soviet Union did produce individuals who were thoroughly capable of entrepreneurship in the field. Such a person was Valentin Gapontsev, whose story comes strikingly close to that of a Silicon Valley garage start-up tale. Gapontsev was a Soviet physicist and specialist in the fundamental physics of light and laser technology who did his graduate work in the 1950s and 1960s at the Lvov

Polytechnical Institute and the Moscow Institute of Physics and Technology. As he later observed, "Although I may have wanted to become an entrepreneur earlier in my career, it would not have been possible in the Soviet Union of that era."[21]

As the Soviet Union was collapsing, in 1990, Gapontsev established a private business in the basement of a small laboratory in the Institute of Radio Engineering in Friazino, near Moscow. In terms of the legal framework customary to Westerners, what Gapontsev was doing was illegal, since he was using state facilities (it was a government institute) for private gain. He did not have the advantage of a private garage. His task, furthermore, was enormously difficult since at that time, the Russian market was frozen. As Gapontsev observed, "It was clear that in Russia they didn't have any real business opportunities, high tech wasn't interesting, and so the market was in the West."

Gapontsev's first contract was with the large Italian telecommunications carrier Italtel, for which he developed a high-power fiber laser amplifier. The Italians were enthusiastic about Gapontsev's products but said they could not accept the risk of working with a small supplier in Russia, and suggested that Gapontsev transfer production to Italy. Gapontsev accepted, and soon developed manufacturing facilities for high-power fiber lasers and amplifiers in both Italy and Germany.

By the year 2000 Gapontsev had built a profitable $52 million company. However, the company, called IPG, ran into the general crisis faced by other telecom companies at the time and was forced to reorganize. Gapontsev once again succeeded in a very difficult business environment but decided that "IPG had to move to the US, because a lot of business is based there."

By 2006 Gapontsev's company, headquartered in Oxford, Massachusetts, had grown to $143 million. The company went public that year. Gapontsev justified the move by saying, "it would be hard to increase our penetration much further without providing customers with the financial transparency and broader awareness that publicly held companies enjoy." Like his countryman Igor Sikorsky almost a century earlier, Gapontsev had found that his ideas could not be commercial successes in his own country.

10   The Exceptions and What They Prove: Software, Space, Nuclear Power

They just beat the pants off us, that's all, and there's no use kidding ourselves about that.
—American astronaut and future senator John Glenn after the successful return of Yuri Gagarin from orbit in 1961

Russia does have some contemporary successes in high technology, especially in the software, space, and nuclear power industries. Let us examine each of them briefly to see what they demonstrate.

Software

The software industry is one in which Russia in recent years has had definite success, although its total software industry is much smaller than that of some other developing countries, such as India. Three different types of software industry have been successful in Russia: offshore programming, packaged software, and software R&D centers in Russia belonging to foreign companies such as Google, Intel, and Samsung. In addition, Russia has a successful search engine, Yandex, which provides services similar to Google's. (I often use it for Russian language searches and find it satisfactory but less sophisticated than Google for advanced searches.) More than half of Russian software exports are in offshore programming.[1]

The best-known software company in Russia is Kaspersky Lab, specializing in antivirus software and winning high praise from major international publications. Moreover, there are hundreds of small software enterprises in Russia, many of them with only a few employees.

Software development plays to Russia's strengths. It is somewhat similar to mathematics in the sense that it relies on the intellectual abilities of a few individuals, often working alone or in a team of two or three people. Software is a creation of the mind, not a material technology. Developing it does not require all the supporting elements that manufacturing a device or a machine would. Russian higher education in mathematics and science is very good. The graduates of the leading higher schools then often go to work in government-financed research institutes or universities and begin to develop software on the side, using in the first stages the computers of their places of employment. Since the capital expenditure to buy a good personal computer is relatively small, they also may work at home. If these individuals are successful, they may associate themselves with several other programmers in a virtual business. A startup emerges.

This fledgling industry may be almost invisible, without a bricks-and-mortar structure. The anonymity of the business gives it relative protection from criminals and bribe-seeking government regulators, who learn about the existence of the business only when it becomes larger and more obvious. It may also escape for a while the scrutiny of the tax authorities, who are notoriously corrupt. In a few cases, such as Kaspersky Lab, by the time the company becomes visible to authorities and criminals it is already large enough and dispersed enough that it can protect itself better than a startup retail establishment or a normal small business operating in an easily identified location.

Even after a full-fledged software enterprise develops, it usually relies more on subcontracts with individual programmers than on regular employees, in that way avoiding paying payroll taxes and benefits. Many scientists and technicians associated with the smaller software enterprises are more like consultants than like employees, selling their services to individual companies who wish to have their payrolls and accounts computerized and to be protected against computer viruses. Criminal elements who specialize in offering "protection" of physical property, such as buildings and storefronts, implying that if the protection money is not paid, they will vandalize the property themselves, have difficulty getting a grasp on such virtual companies. Just where is the physical property? Even the larger software companies do not present as many entry points for corruption as a normal factory or retail business does since much of the activity of these businesses occurs in dispersed places, such as private apartments and academic labs.

The history of Kaspersky Lab fits this model closely. Today it is a large company, with annual revenues well over half a billion dollars. It is the only Russian firm that holds a place in the world's top one hundred software companies by revenue. Its cofounder and principal personality, Eugene Kaspersky, was strongly attracted to mathematics at an early age. In 1987 he graduated from the Institute of Cryptography, Telecommunications, and Computer Science, an organization jointly sponsored by the KGB and the Soviet defense establishment. After graduation he worked for a while designing antivirus protection programs for several different Ukrainian and Russian companies, earning about $100 a month. He formed a three-person team that became outstanding in antivirus software design.

In 1994 a German university noticed his work and called his toolkit perhaps the best antivirus scanner in the world. Kaspersky and his team soon started receiving licensing requests from European and American computer companies, and in 1997 he cofounded Kaspersky Lab with his wife, Natalya, who also had a technical education and background. During the next few years the company still operated below the radar of the authorities, but eventually emerged as a major company. Kaspersky himself became a multimillionaire. By this time he could afford to protect himself, as soon became necessary.

Although Kaspersky Lab has a Moscow office, it is still unusually dispersed. Most of its business is done abroad, outside Russia. Many members of its "Research and Analysis Team" live and work outside Russia, in eleven different countries, including Germany, the UK, France, Sweden, the United States, and Japan. Even within Russia many of Kaspersky Lab's employees live in cities other than Moscow, such as St. Petersburg and Novosibirsk, and some are still graduate students working for advanced degrees in local universities at the same time as they do work for Kaspersky.

Eventually criminal elements noticed Kaspersky, saw that he had become a very wealthy man, and decided they wanted their slice. But gaining leverage over such a scattered operation was not so easy. They chose a personal approach. On April 29, 2011, criminals kidnapped Kaspersky's twenty-year-old son, Ivan, a student in computer science at Moscow University, while he was on his way to work. The kidnappers then telephoned Eugene Kaspersky in London, where he was on a business trip, and demanded a ransom of $4.4 million for the return of his son. Eugene Kaspersky immediately flew back to Moscow and organized, together with the police, a trap for the kidnappers, promising them the ransom money. The police caught the intermediary picking up the money and traced him back to the main criminals. Five of them were arrested. No ransom was paid. It was an unusual victory for justice in Moscow, no doubt aided by the fact that Eugene Kaspersky already knew a thing or two about criminality from his long experience in computer security.

Even today, at a time when Kaspersky Lab products can be purchased all over the world, the company maintains a strikingly low profile in Russia. I recently went to its main office at 10/11 1st Volokolamsky Proezd in Moscow. Nowhere could I see the words "Kaspersky Lab" advertising the presence of the company. Its headquarters are located in a secure area, a nondescript, unlabeled, and guarded building that is a business park for other companies seeking similar anonymity. The closest I could get to the Kaspersky offices was the front door of the business park, where I asked the guard if I could speak to the receptionist for Kaspersky Lab. "No," I was told by the guard, "one must have permission in advance to enter the premises." I found out, by looking at the small telephone directory on the guard's desk, that Kaspersky Lab occupies four floors of the building. The telephone directory was the only place I saw the name of the company whose world headquarters I was trying to visit.

Space Technology

Russia's success in the software industry is particularly interesting and enticing because it is an example of a post-Soviet achievement. Although Russian software certainly builds on the strength of Soviet education and science, its actual existence is almost entirely a post-Soviet phenomenon. Other areas in which Russia today has high-technology strength, such as space technology and nuclear power, are largely remnants of earlier Soviet achievements. The Russians have not done much new in these latter areas beyond what their Soviet predecessors did, but they nonetheless still have some genuine strengths.

The Soviet Union was obviously a pioneer in space exploration. After all, it launched the world's first artificial satellite and the first human being into space.[2] Its rockets established a strong reputation for power and reliability (although, like the United States, it had its spectacular failures as well). I was in Moscow as a student on April 12, 1961, when Yuri Gagarin first circled the earth, and later I briefly met him.[3] I remember Gagarin's achievement as the apogee of Soviet self-confidence. Soviet science and technology were thought to be the best in the world; the enthusiasm of the thousands of celebrants in the streets with whom I mingled was heartfelt and genuine. In the history of fits and starts of Russian technology, this was undoubtedly its most glorious start.

After the retirement of the space shuttle program, the United States was for a while dependent on Russia for transportation to the International Space Station. The Soyuz rocket used for that purpose is, in terms of design, over forty years old, but it is the most often used and the most reliable such rocket in history. It has been launched more than 1,700 times.

Another example of the strength of Soviet rocketry can be found in the use by American space companies of Soviet rockets after the dissolution of the Soviet Union and the end of the Cold War.[4] In the 1960s and early 1970s the Soviet Union produced large numbers of the NK-33 rocket engine, many of which ended up mothballed in storage. In the 1990s an American developer of rocket engines, Aerojet, purchased thirty-six of the NK-33s for use in commercial satellite launches. The Soviet-designed rocket engines were better than anything that Aerojet designed itself.

The excellence of these Soviet rocket engines demonstrates that in technology projects to which the Soviet government assigned the highest priorities and showered with bountiful resources, the results were often impressive. Space technology and atomic weapons were given almost unlimited resources, in terms of both money and talented personnel. Questions of cost-effectiveness rarely arose.

Now in both Russia and the United States, as well as in other countries, that period is looking increasingly dated. An unmanned Russian Mars mission in November 2011 failed to escape Earth's orbit.[5] The United States is trying to reduce the costs of its space program by privatizing segments of it, by encouraging private companies to develop low-orbit launch vehicles that can be successful commercially. NASA has already worked toward this goal with the companies Rocketplane Kistler, SpaceX, Orbital Sciences Corporation, and Boeing. Russia stands in danger of losing its earlier eminence in space if it does not do something similar. One is reminded of the early history of computers. When computers were largely government financed, in the 1940s through the early 1960s, the Soviet Union competed successfully in computers with Western nations. But when computers became commercial commodities and market competition became intense, the Russians fell behind. While there will always be a role for governments in space exploration, some of the innovations in the field are likely to come from entrepreneurial sources of a type Russia has so far failed to develop. It is worth noticing that as of 2012, Russia's share in the largest commercial section of space technology, satellite communications and telecommunications (over $100 billion), was less than 1 percent.[6] Even in Russia's vaunted space technology sector the pattern seen so often in this book holds.

Nuclear Technology

Russia is a powerful world competitor in nuclear technology. Historically, its strength in this area derives from the Soviet nuclear weapons program, but in the post-Soviet period the Russian government has continued to promote nuclear power as a technology in which it enjoys export advantages.[7] Its state nuclear energy agency, Rosatom, has formed a joint-stock subsidiary corporation, Atomenergoprom, which is the largest company in the world in terms of exports of nuclear power plants. This company, with its various subsidiaries, offers services at all stages of the nuclear power industry, from uranium mining and fuel manufacture, to nuclear reactor design and production, to sales of nuclear electrical power plants. It also offers important services in uranium conversion and enrichment, possessing the world's largest uranium enrichment capacity. For uranium enrichment it uses gas centrifuge technology, a method that is less costly than the gas diffusion technology often used in Europe and America. In uranium enrichment Russia supplies Western Europe with one-third of its nuclear reactor fuel.

Thus, Russia enjoys at the present time genuine advantages in the world market in nuclear power technology. Worries about safety, rooted in the Chernobyl disaster of 1986, continue to hamper sales in the West and are undoubtedly a reason why Russia has been most successful in selling nuclear power plants to China, Iran, and India, where local government support has been strong.

However, the future of Russian export of nuclear power technology is uncertain. After the 2011 Fukushima nuclear power disaster governments are rethinking their policies on nuclear power. Germany has decided to phase it out. France, a remaining strong proponent of nuclear power, relies mostly on its own technology, although it utilizes Russian enrichment facilities. After Fukushima there is an even stronger emphasis than before on finding safer nuclear power technologies, with hopes for radically different emphases, such as mini-power stations. New technologies for producing nuclear fuel, such as laser isotope separation, are also emerging. Russia will be hard-pressed to keep up with all these new developments, but it continues to enjoy an impressive nuclear research and development establishment.

II What Are the Causes of the Problem?

How does one explain the pattern of impressive technological invention in Russia followed, again and again, by failure to develop and sustain that invention as a true innovation? We have seen that pattern in the arms industry, where Tula in the seventeenth century and again in the early nineteenth had one of the most impressive weapons-producing centers in the world; in railways, where in 1847 American engineers said the Alexandrovskii works in St. Petersburg were the most modern they had ever seen; in the electrical industry, where London and Paris in the 1870s were dazzled by the "Russian lamps" that illuminated their fashionable avenues; in aviation, where Russians before World War I built remarkable passenger planes; in Soviet industrialization, during which the world's largest steel mills and hydroelectric plants were constructed; in biology, where Russians in the 1920s and early 1930s were leaders in the "new synthesis" of evolutionary biology and the new genetics that would lead elsewhere to spectacular technological development; in the semiconductor industry, where Russian engineers anticipated, in some cases, the world industry by almost a generation; in computers, where Russia built a very early electronic computer and one of the fastest in the world; in lasers, where Russians won the attention of the world with their pioneering research, garnering Nobel Prizes; and in space research, where Russians launched into orbit the first artificial satellite and the first human being.

In all these cases the early promise was not fulfilled. Instead, we have witnessed what can only be called massive failure in sustaining earlier excellence. Today, Russia is a minor player in the world high-technology market. Once again, Russian leaders are forced to repeat what their predecessors from Peter the Great on have proclaimed: Russia must modernize its industry.

No other country in the world displays this pattern of intellectual and artistic excellence and technological weakness to the same degree as Russia. It is a phenomenon of world significance that calls for explanation. This pattern has determined the fate of Russia as a nation, not only consigning it to backwardness but also providing the excuse for authoritarian government, from Peter the Great to Stalin to Putin. Russia's retardation in sustaining technology is thus not just a chapter in the history of technology; it is an important key to its political and social evolution because it is one of the important reasons why its leaders can ignore true democracy and call again today for forced-draft modernization through political compulsion, not realizing that by so doing they only reinforce the fateful pattern.

Russia is a graphic illustration of the general principle that technology, once introduced, does not automatically spread and become indigenous. Sustaining technology requires a society that supports and inspires it, a society in which innovation becomes a natural development. Russia to the present day has failed to do that, and as a result, once again, in the second decade of the twentieth-first century, recent and current Russian leaders like Medvedev and Putin are calling for technological modernization. It is the same message preached by many of their predecessors—Gorbachev, Brezhnev, Khrushchev, Stalin, Lenin, Alexander II, Catherine the Great, Peter the Great.

What is the place of Russia in the history of technology worldwide? Failure in technology is certainly not a specifically Russian phenomenon. A large Western literature exists on technological failure.[1] Much of that literature addresses notorious examples of technologies that at first looked brilliant but later were not successes, such as Charles Babbage's early computers (1847–1849), Sony's Betamax (1975), Polaroid's Polavision (1977), Apple's Newton (1993), or the Segway PT (2001). For the historically minded, Leonardo da Vinci's "helicopter" or "tank" (1480s) might be named. Brazilians often still contend that the first true aviator was Alberto Santos-Dumont (early 1900s) and not the Wright brothers. In each of these cases, brilliance in design did not result in commercial success.

Is Russia's pattern of technological creativity followed by practical failure just another example of a worldwide phenomenon? No. Russia's repeated failures to sustain technology over three centuries catapult it to an unfortunate special status. No other country has such a consistent record, both brilliant and dismal. Instead of customary explanations for technological failures based on the characteristics of a specific device, such as "ahead of its time" or "too expensive" or "lack of investment support" or "flawed in design" or "poorly marketed," in Russia's case we must talk about larger societal obstacles to technological success. Only such a broad approach can encompass failures over such a long time.

The explanations for this pattern can be grouped into several large categories: attitudinal, political, social, economic, legal, and organizational. Some parts of these explanations, such as the lack of adequate and effective legislation for innovations, are fairly obvious and easy to explain; others, such as the attitudinal one, are elusive but nonetheless of great importance.

11 The Attitudinal Question

We look upon a scientist as a disinterested person who does everything for the good of humankind. An entrepreneur is a member of the bourgeoisie who takes advantage of other people.
—Russian scientist in 2010, responding to a survey about attitudes toward science and technology

One of the factors limiting Russian efforts in technology is attitudinal. It is difficult to analyze, cannot be measured in economic terms, and is even somewhat speculative, yet it may be the most important of all. Russians have never to the present day fully adopted the modern view that making money from technological innovation is an honorable, decent, and admirable thing to do. In the nineteenth century, throughout the Soviet period, and still today, *biznes*, or "business," has often been seen by Russians as a disreputable activity. Intellectuals (*intelligenty*) in particular saw (and often still see) commerce as below their dignity. In recent post-Soviet years the connection of successful businesspeople, especially the oligarchs, with corruption has only deepened suspicion of business operations.

We should recognize that many Russians simply do not want a Western, liberal, competitive, open-market system. They have often desired to go their own way, maintaining they are pursuing "higher values." Russia's only living Nobel laureate in science, the physicist Zhores Alferov, told me in December 2011 that he considered the fall of the Soviet Union "a great political, moral, and, first of all, economic tragedy." Alferov is cochair of the Scientific Advisory Council of Skolkovo, Russia's effort to duplicate Silicon Valley. The head of the Russian government, Vladimir Putin, similarly called the end of the Soviet Union the "greatest geopolitical tragedy of the twentieth century." Attitudes such as these hinder Russia's entry into today's global high-technology economy.

Over the past fifty years I have made more than one hundred trips to the Soviet Union and Russia for periods of time totaling several years, and have talked to several thousand Russian scientists, engineers, and science and engineering students, in both informal and formal settings. In 2005–2013 alone I visited approximately sixty universities and research institutes all over the country, ranging from St. Petersburg in the west to Moscow at the center of political power, multiple cities in the industrial Urals region, and throughout Siberia to such places as Tomsk, Novosibirsk, Krasnoyarsk, and Vladivostok. As a former engineer myself, I was attracted to conversations with engineers and scientists. The fact that I had studied at Moscow University helped in initiating our encounters. During all these travels I was actively comparing attitudes I met in Russia with those I knew in my home institution, MIT, where I was a professor.

When I asked science and engineering undergraduates at MIT what their professional goals were, I of course received a variety of answers, but a strikingly frequent one went something like this: "I would like to start my own high-tech company and make a success of it. If I cannot be the next Bill Gates or Steve Jobs, then at least I want to create a company sufficiently valuable that I can sell it for a high price to one of the existing major companies in the U.S. Then I will try to find an idea for another start-up."

I can truthfully say I have never heard such an answer from a Russian student. Russian students—and working scientists and engineers—just do not think that way, although an enormous effort is currently being made in Russia to change attitudes on this subject (more on that in the final section, "Can Russia Overcome Its Problem Today?").

I have always wanted to have more objective information to support or contradict my anecdotal, personal experiences. Such information is difficult to find. However, some is available. In 2010, scholars at the European University in St. Petersburg conducted a survey of Russian scientists and engineers asking them about their attitudes toward their work, and they also drew on a larger sociological study on professional attitudes conducted by the University of Magdeburg.[1] One wishes that the total number of in-depth interviews with scientists—only several dozen—had been greater, but the studies still illustrate frames of mind that are clarifying.

One respondent replied,

There are no models in the consciousness of [Russian] people of a successful scientist-entrepreneur. We look upon a scientist as a disinterested person who does everything for the good of humankind. An entrepreneur is a member of the bourgeoisie who takes advantage of other people." [The respondent was forty-one years old in 2010, meaning he was twenty-one at the time of the collapse of the Soviet Union]

Another respondent said,

We must talk about our inability to commercialize our own products. This is not a misfortune of the Soviet Union, it is a misfortune of the Russian mentality in general. . . . To our regret, up to the present day society does not have a very positive attitude toward the commercialization of scientific ideas.

Yet another Russian scientist, one with more than fifty international patents, said,

You know, I do not have a commercial bent! I have an idea and my goal is to realize it. And when I manage to do that, when I get a good result, that means I will publish it or maybe patent it. And then I am content. To go further is not my affair—to try to apply all that in business requires so much work of a sort that is not interesting to me. And as a result other people [in other countries] simply rob from us. Right now several of my innovations are being shamelessly used by companies in, for example, China and Israel.

And another young scientist said,

We do not have an innovation culture—no experience, no traditions. Our scientists, they are all still Soviet in their attitudes, for them "business is something dirty." Our scientific culture is practically untouched by the business entrepreneurial spirit.[2]

How does one explain this negative attitude toward the application of science to commercial technology among many Russian scientists? The answer can be found in an unusual confluence of obsolescent ideas present in premodern European history and the specifics of recent Russian history. Russia has suffered from both an ancient general disease and a modern unique one.

The ancient general disease can be seen in European history. In premodern Europe, making money from "trade" was often looked down upon. The monarchy, the nobility, and the church all derived their status from inherent rights, not achieved or earned ones. The monarchy ruled by divine right, not because of its ability or achievement; the nobility enjoyed special status because of its ancestry and its service role as the defender of the lord or the nation. The church occupied itself with the spiritual world and the religious justification of the existing moral order. All this began to change in Western Europe at the end of the seventeenth century, first in the Netherlands, then in England, then in North America, and then in the rest of

Western Europe.[3] A novel idea took root: it was possible to be a decent, respectable, even admirable citizen while making money through one's cleverness in manufacturing and distributing products or offering services. To some degree this idea was linked to Protestantism and nascent capitalism (the Weber thesis), but it had a status of its own and developed in some areas without these congenial companions.

Russia was affected by this idea much later than most of Western Europe. To the end of the tsarist empire, the strengths of the monarchy, the nobility, and the church were greater than those of a rising bourgeoisie. Prestige was linked to strength, and merchants and entrepreneurs did not have high social standing. Protestantism was largely absent,[4] and if capitalism had come to Russia by the late nineteenth century it was still incomplete, even unjustified in the views of many social critics, both those still attached to romantic ideas of the peasantry and those affected by radical, Marxist ideas coming from Western Europe.[5] When successful businessmen or financiers were Jewish, another cause of hostility to *biznes* was anti-Semitism. By the nineteenth century Russia had a small but capable community of scientists, but most of them occupied themselves with the world of the mind, little connected to practical activity (there were a few exceptions, such as the outstanding chemist Dmitry Mendeleev).[6] Russia's strengths in abstract mathematics (for example, non-Euclidean geometry) and in physics and chemistry (but not industry based on them) were already established.

To this premodern set of mind, still prevalent in Russia to the end of the tsarist period, was added, in the last decades of the old regime and throughout the seventy years of the new Soviet one, an influx of radical ideas highly critical of capitalism, competition, and private endeavor. The Marxist revolutionaries who took over Russia in 1917 were modernizers, but they saw the state, especially a state planning apparatus, as the key to modernity, not the efforts of individual entrepreneurs. Thus, the idea of the innovator making money from creativity, an idea already underdeveloped in tsarist Russia compared to much of the rest of Europe, became in Soviet Russia a disrespectable, even immoral one. The major Soviet encyclopedia defined the "bourgeoisie" as "the ruling class in capitalist societies living by exploiting the labor of hired workers." Entrepreneurs need to hire workers. In 2006 a new encyclopedia removed the pejorative phrase, saying merely that the "bourgeoisie" was a "social class owning capital," but the older reference work was still being used widely, and even the new one was hardly positive about the role of the bourgeoisie.[7]

To Russian scientists being supported by the government in state research organizations such as the institutes of the Academy of Sciences, "the Soviet ideology that condemned private entrepreneurial activity was not an unpleasant one." It gave them a status not unlike the church in premodern Europe: they lived in a world of ideas, and if, in contrast to the church, rewards came from brilliance, those rewards, similar to those in the church, were not connected to utility.

"Even if some scientists were critical of the political controls they saw around them in the Soviet Union, those who rose to the top in their research organizations valued the special, unburdened status that the government gave them and the access to special groceries, hospitals, rest homes, and travel and other perquisites denied to many others. The remarkable privileges that leading scientists in the Soviet Union enjoyed, regardless of their actual assistance to the economy, helps to explain why, when the Soviet Union began to collapse, top administrators of the science establishment were among the most ardent defenders of the old order.[8] The president of the Soviet Academy of Sciences, for example, stoutly defended the old Soviet system even as it imploded for lack of support elsewhere. And some older scientists still today look upon their status in Soviet times with nostalgia. They are reluctant to face a world of economic competition.

There are some signs lately of a nascent attitudinal change in Russia toward the commercialization of technology. In Russian business schools, schools of management, economics departments, and in government statements, many calls for the "commercialization of technology" are being made. Start-ups, business incubators, science parks, technology platforms, and "clusters" for innovation are springing up in many areas. These efforts are weaker among scientists than among management specialists; they are still rare in the Russian Academy of Sciences, and also not very strong in university science departments. But they are beginning.

Several American foundations are helping this transition to occur. The U.S. Civilian Research and Development Foundation (CRDF), based in Arlington, Virginia, has for many years pushed for this change through the technology transfer offices it has helped establish in many Russian universities and through special programs, such as the First Steps to Market Program. The Eureka (*Evrika*) Program, launched in 2011, has pushed the same agenda and has established partnerships between American and Russian universities (participating institutions include Purdue University, the

University of Maryland, the University of California at Los Angeles, Nizhny Novgorod State University, and St. Petersburg University of Information Technologies, Mechanics, and Optics [ITMO]). During a recent visit to the latter Russian university I heard a young graduate student describe the start-up company he wished to establish for the electronic avoidance of traffic jams, a major problem in large cities such as Moscow. The Eureka Program is sponsored by the U.S. Russia Foundation for Economic Advancement and the Rule of Law (USRF), the American Councils for International Education, and the New Eurasia Foundation. And, of course, yet another important organization in this commercialization effort is the Skolkovo Foundation.

A major weakness of all these efforts is that they do not address the major societal reforms that are necessary for strong technological development to occur. As Daron Acemoglu and James A. Robinson maintain in their recent book, *Why Nations Fail,* inclusive social and political institutions are strong factors in fostering economic development.[9] In Russia, where one political faction rules, where independent and nongovernmental institutions are restricted, where political critics are silenced, and where the mass media are controlled, the development of such attitudes and institutions is very difficult. As Yegor Gaidar, prime minister of the Russian government under Yeltsin, recently observed, "The Russian political elite wanted to borrow military and production technology and not the European institutions on which the achievements of Western Europe were based."[10]

In concluding this "attitudinal" chapter I should recognize that several people who have read drafts of this book have asked, "Why do you use the word 'attitudinal' rather than 'cultural'? Is not this a cultural problem?" Yes, I suppose I could use "cultural," but the word *culture* seems a bit too large and too amorphous (in the expansive sense in which anthropologists use the term) to convey what I have seen again and again in conversations with Russian scientists: their critical attitudes toward business applications of their work. I wanted to sharpen the focus of my observations by using the term "attitudinal" because it is precisely the attitudes of these scientists that have struck my attention, much more so than broad characteristics of Russian culture as a whole. It is a question of opinions about applied science and technology (as opposed to fundamental science or abstract thought) that I wished to emphasize. But I would accept "cultural" if my use of the term were correctly understood.

Aleksandr Lodygin (1847–1923), one of the inventors of the incandescent lightbulb.

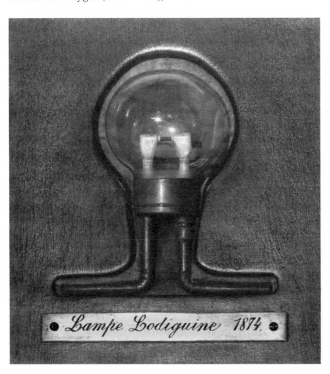

Aleksandr Lodygin's electric lamp of 1874.

Pavel Yablochkov (1847–1894), who first electrically illuminated the streets of Paris and London.

Aleksei Krylov (1863–1945), a talented naval engineer and mathematician who in 1904 invented a machine for the solution of differential equations.

Igor Sikorsky (1889–1972) with his prizewinning S-6 plane, 1912.

Tsar Nicholas II inspecting Igor Sikorsky's four-engine airplane, 1913.

Aleksandr Popov (1859–1906), considered by Russians to be the inventor of the radio.

Georgii Karpechenko (1889–1941), first creator of a new species by polyploid speciation. Executed on July 28, 1941.

Valentin Fabrikant (1907–1991), who first developed the theory of the laser in his doctoral dissertation of 1939.

A Soviet-era poster of ca. 1930 depicting the industrial and agricultural bounty that Soviet socialism would bring.

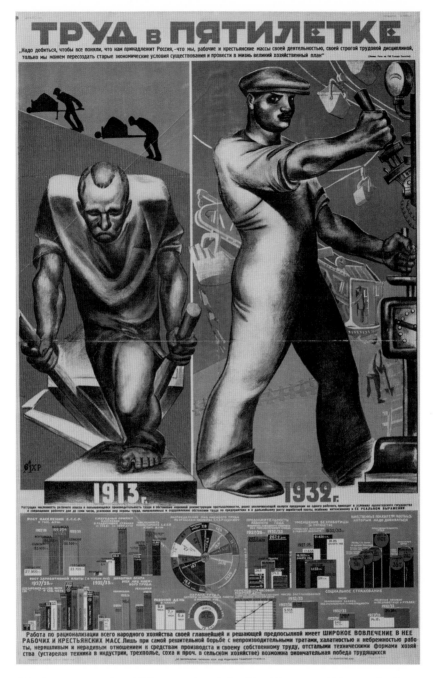

A Soviet-era poster of ca. 1932 contrasting the backbreaking manual labor predominant in Russia in 1913 with the mechanized labor of 1932.

A Soviet-era poster quoting Stalin's call to "catch up with and surpass" Western countries in industrial production and contrasting the enslaved proletariat of the West with liberated Soviet workers.

A Soviet-era poster depicting how industrial production of tractors would transform the countryside.

A Soviet-era poster calling for the completion of the five-year industrial plan in four years.

The *Maxim Gorky*, the world's largest airplane in 1935, flying over Red Square in a demonstration. Shortly after this photograph was taken one of the small planes crashed into the large plane, killing 48 people.

Alexander Prokhorov, co-winner of the Nobel Prize in Physics in 1964 for the development of the laser, demonstrating his instrument.

Sergei Korolev, lead spacecraft designer, Igor Kurchatov, director of the Soviet atomic bomb project, and Mstislav Keldysh, chief theoretician of the space program. This Soviet-era photograph is surely a montage, judging by the border between Keldysh and Kurchatov.

Yuri Gagarin, first man in space, with Sergei Korolev, head of the Soviet space program.

Mikhail Kalashnikov with his invention, the AK-47.

12 The Political Order

Citizens recruited by international organizations acting against the country's interests will also be considered traitors.
—The official newspaper *Rossiiskaia gazeta,* commenting on Vladimir Putin's signing of a new law on treason, November 19, 2012

The political problem has been, in a word, one of authoritarianism. The tsars, the leaders of the Communist Party, and now the leaders of post-Soviet "sovereign democracy" (which is not democracy at all) have determined the policies that govern technological development, often ignoring market forces and "best practices" that, in at least many instances, governed the development of technology elsewhere. Of course, Russia is not the only country in the world to follow mistaken policies in technological development; similar instances of unfortunate policies can be found in all industrialized countries in the world, including Germany, the United States, and Japan, but Russia stands out for the extreme character and frequency of occurrence of such policies.

"Political authoritarianism" is not the same thing as "state control." Although central governments do not have a good record in predicting technologies, a centralized government in a state that is democratic can play a useful role in promoting technology. France's government is highly centralized but it has been successful in at least some of its state-led megaprojects, such as the TGV (high-speed railroads) and nuclear power. In others, such as its attempt to create its own Silicon Valley in Sophia Antipolis on the Riviera, it has been less successful. China is a special case in that it is both centralized and nondemocratic. China is, in fact, the greatest challenge to the basic thesis of this book, that technology is most creative and successful in a democratic, law-governed society.[1] Much will depend on the future, but China so far has been much more successful in achieving economic growth than it has in creating its own novel high technologies.

From the very beginning of our story we see that the rulers of Russia, the tsars, were more interested in technology that increased military might or made deep impressions on observers or favored elites than they were in improving the Russian economy. This pattern has continued to the present. Militarization has distorted Russia's technological development. Peter the Great gave the modernization of Russia a tremendous push, but his primary motivation was to increase Russia's standing as a European military power, not to improve the standard of living of his people. He modernized the armaments factories, importing the latest technologies from Western Europe, and these efforts served him well in his several wars. After his reign his successors continued to support arms industries, but only sporadically, at moments when the obsolescence of previous systems of production resulted in military defeats, as in the Crimean War, which then spurred a sudden effort to catch up. In addition, the tsars and nobles who controlled the industry desired "ornamentation" or "presentation" arms for their personal use. They rewarded those who could produce such impressive items. As a result, some of the most talented of the gunsmiths devoted their efforts not to innovation or the mass production of lethal weapons for winning wars but to crafting beautifully decorated arms, damascened and inlaid with precious metals, for the pleasure of their royal and noble superiors.

A similar desire to use technology primarily for its presentation effects can be seen in Khrushchev's and Brezhnev's demands that Soviet cosmonauts make spectacular flights on important political dates, such as summit meetings or the anniversary of the Russian Revolution. Perhaps the most extreme of these requests by Khrushchev for presentation technology (similar to the tsars of old) came in 1963, when he demanded that the Soviet Union launch three men in one space vehicle before the United States succeeded in launching two in a single capsule. To carry out that order, Khrushchev's talented designer Sergei Korolev had to choose cosmonauts of small stature, ask them to drop the precaution of wearing bulky space suits, and pack them in a small sphere so tightly that they were arranged around each other like pretzels. But the effort succeeded, and on time.[2]

Political factors severely hindered innovating engineers and scientists throughout Russian history. The pioneering electrical engineer Pavel Yablochkov, who first illuminated the central avenues of Paris, lived in that city long enough to become friends with Russian political exiles residing there, including the Russian translator of Karl Marx's *Das Kapital*. That association was enough to ensure the suspicion of the tsarist police and their hounding of Yablochkov when he returned to Russia and tried unsuccessfully to establish an electrical company there.

Such examples of politics interfering with innovation could be easily multiplied in the Soviet period. The aeronautical engineer Igor Sikorsky fled repressive politics and emigrated to the United States, where he established his own successful company. A pathfinder in the development of television, Vladimir Zworykin, escaped from civil war in Russia in 1919, also to the United States, where he worked for Westinghouse and RCA. Oleg Losev was a great pioneer in the development of semiconductors and diodes, but because of his questionable background (noble birth and private economic activity in the 1920s) he was limited in his ability to introduce his innovations into the economy. The chemical engineer Vladimir Ipatieff was driven out of the Soviet Union and also fled to the United States, where he developed gasoline-refining methods for the Sun Oil Company. The outstanding geneticist Theodosius Dobzhansky was similarly forced to emigrate, and became a leading faculty member of Rockefeller University. The list goes on and on. Even today, Russia suffers from brain drain and capital flight as scientists and investors seek better conditions abroad. Many Western European and American universities have on their faculties (especially in the mathematics and physics departments) former Russian citizens who left their homeland for political reasons.

The Soviet Union achieved an impressive industrialization, but the effort was severely distorted by politics. Stalin, for example, emphasized aviation, but wanted international laurels more than he wanted innovation or efficiency. By 1938, Stalin could claim sixty-two world records for his pilots, whom he called his "falcons," including the longest, highest, and fastest flights anywhere. His pilots and aviation designers went to great lengths to achieve these records, often cutting corners and building planes that were economically inefficient and incapable of competing in world markets even though they were capable of establishing world records. But the Soviet Union was an isolated country, not a major participant in world trade, so the inability of its airplanes to compete with others in economic terms was not an important factor so long as the Soviet economy was a protected one. But the habits of design that became a part of the Soviet aviation industry during that period would place it at a disadvantage whenever it became exposed to the world market, as it later did. Today the Russian aviation industry has great difficulty in the international market.

The entire Soviet industrialization effort was seriously distorted by political considerations. Driven by an ideological imperative to "conquer nature," Soviet industrial planners built giant cities and factories in remote and frigid locations that were inherently inefficient from energy and transportation standpoints. As a result, Russia now faces the staggering task of rebuilding and relocating much of its industrial infrastructure if it hopes to compete in world markets, where cost efficiencies are often determining.

In the post-Soviet period the Russian government has tried to modernize in a more rational way and has emphasized high-technology projects, such as the Skolkovo innovation city, to which many foreign companies and universities, especially MIT, have been attracted. However, political problems have not disappeared. For example, in November 2012 Russia's legislature passed, and Vladimir Putin signed, a new law that broadens the actions that can attract treason charges to include giving "financial, material, *technical* [my emphasis], consultative or other aid" to foreign organizations.[3] No one knows how this law will be interpreted, or just how selectively it will be imposed. Nonetheless, a Russian scientist or engineer working together with foreign counterparts on research that may have dual-use (both civilian and military) applications has reason to worry. This anxiety inhibits free research activity and, ultimately, modernization.

13   Social Barriers

Russians over 14 currently have to notify the federal migration agencies in person if they move to another place in the country where they intend to stay for more than 90 days.

—Russian International News Agency (RIA Novosti), November 11, 2010

Innovation is increased in society by social and geographic mobility, by the ability to live where one wants, seeking the best location and resources for certain types of technological and economic activity. Silicon Valley in California, the Route 128 high-technology corridor around Boston, and similar spots in Israel, Great Britain, and other democratic countries did not result from people being directed by their governments to go there but instead flourished from the decisions of talented people that those were the best places to be for what they wanted to do. Of course, democratic governments help foster innovation in certain locations, often offering tax and other inducements if industries establish themselves there. Furthermore, industries and military installations often transfer their employees as a part of their service obligations. As is the case with many other comparisons, the differences between authoritarian societies and democratic ones are often matters of degree, but these differences are nonetheless very important. In democratic industrialized societies much of the movement of the most innovative people has been voluntary.

From the times of Peter the Great to the present day, the Russian government has tried to control mobility, with limitation on innovation a consequence. The system of serfdom curtailed the independence of the artisans of Tula in the arms industry and the entrepreneurial activities of innovators. Attempts to leave the enterprise sometimes resulted in severe corporal punishment, as we saw in the case of the Tula worker Ivan Silin, subjected to lethal lashing for leaving his place of work. The difference between George Stephenson in England and Miron Cherepanov in Russia is striking: both came from simple mining families and both developed working locomotives, but Stephenson was able to attract private investors, move to a place where he could find workers, and establish his own company, whereas Cherepanov, a serf, could not even think of such activity and was unable to interest his noble owners in developing his innovations.

Even after the abolition of serfdom, Russian citizens were much more bound to place of residence and work than the citizens of Western industrializing societies. The controls were both internal and external. Going abroad for study was strictly regulated, as was the mobility of "undesirable" people such as political dissidents, Jews, or Roma (gypsies). The enforcement of these controls varied from time to time, and sometimes a clever person could evade them, often through bribes or deception, but the rules were nonetheless a constant consideration. The great mathematician Sofia Kovalevskaia managed to get around these regulations and receive a fine graduate education in Switzerland by taking the radical step of contracting a fictitious marriage with another Russian who had permission to travel abroad. (The two actually came to like each other.) Because residents were required to register their place of residence with the police, "suspicious" people could be followed and their freedom restricted.

The tsarist government established an internal passport regime, called *propiska*, that tied Russian subjects to places of residence. These official residences were listed in their passports, which they were required to carry and which police could demand to inspect at any time. If Russian subjects wished to travel, they had to register with the police to do so.

When the first major railway was built, between St. Petersburg and Moscow, the question of how to handle this new means of mobility was a serious challenge to the government, and at first travelers were required to arrive at the train stations hours in advance in order to register their travel with the police. Later, these rules were softened, but to the end of the empire a person desiring to leave his or her place of residence for more than a certain number of days was required to get police permission. The government could decide when to relax or modify these controls, and it did so in an attempt to manage urbanization and foster modernization of the sort it favored, such as the building of the Trans-Siberian Railway.

At the time of the revolutions of 1917 the system of *propiska*, or mandatory residency permits, was abolished, but it was reinstituted in 1932, and vestiges of the system remain even today, in post-Soviet Russia. During Soviet times all citizens were required to have internal passports in which their residences were listed. Different types of *propiska* existed (such as "permanent," "temporary," and "service"), but mobility was controlled. Often a person could retain residency in a desirable place (such as Moscow or Leningrad) only so long as he or she continued to work for a certain economic entity (factory, ministry), which also often provided the apartment in which the person lived. The thought of quitting one's place of work in order to do something more innovative in a different place or for a different employer was usually out of the question. One would lose both one's apartment and the right to live in that city.

Foreigners were strictly controlled (when I studied in Moscow I was given an internal passport, a *vid na zhitel'stvo*, in which my residency was listed, and I was told I could not travel more than 45 kilometers from it without police permission). Dissidents had little chance of receiving official permission to move to a different place. Ordinary citizens could not get a job, marry, move to a different residence, access health care, get a state education, or receive a pension without presenting their passports and, often, "work books." Sometimes bribes and evasion helped one accommodate to the regulations (again, fictitious marriages, as in tsarist times, might permit a person to move to a desirable place), but they were always there.

Under these kinds of restrictions a talented engineer who came up with an innovative idea leading to a new marketable product knew that unless he or she could convince his or her current employer to produce the item, there was no chance for its development. If we consider where successful innovations come from in other societies, the harmful effects of such regulations are obvious. In the West, some of the most innovative products in fields such as electronics, lasers, software, and computers have come from restless employees of large firms who left to start their own companies. Many failed, but a few succeeded brilliantly, transforming entire industries. Both Intel and Apple Computer, today giant companies, were established in this way (Gordon Moore left Fairchild Semiconductor; Steve Jobs left Atari).

The Soviet government used these residency requirements to foster the projects it favored. For example, when the government wanted to establish a new "science city" far away in Siberia, in Novosibirsk, in the 1950s, it knew that convincing distinguished scientists from privileged cities such as Moscow and Leningrad to move there would not be simple. It offered an unusual attraction tied to residence permits: leading scientists who agreed to help establish research laboratories in Novosibirsk could retain their residency permits in their home cities, so that they could have apartments in both Moscow and Novosibirsk (actually, some even had "cottages" in Novosibirsk, extremely rare in the Soviet Union). "Closed science cities" where access and residency were strictly controlled were established all over the Soviet Union, and a few remain even today. In Soviet times these places often had favored supply and distribution systems, offering researchers privileges not available to normal citizens. The *propiska* system could thus be used as both a stick and a carrot. The new innovation city of Skolkovo currently being created near Moscow is being promised similar privileges, even though the word *propiska* is no longer officially used.

In post-Soviet Russia the *propiska* system has never entirely gone away. Officially it was abolished, but a scaled-down version has been retained. Now the official word is "registration" (*registratsiia*), but in popular parlance the term is still *propiska*. Citizens of Russia should be "registered" if they live in the same place for ninety days. A place of permanent residence is still written in the internal passport, and it is an administrative offense to live in a dwelling without registration. Foreigners are required to fill out a "migration control card" and to register with the police within a short time after they check in to a hotel.

The system of residency control in Russia today is not nearly as restrictive as it was in Soviet and tsarist times, but geographic mobility is still much less than in Western industrialized countries. The attitudes of Russian government officials toward geographic mobility are similar to those of Soviet officials. People can still be stopped on the street by the police and asked for their passports and residence registration, and if they are in the "wrong" place they can be sent away, especially if they are members of a group that is the object of prejudice and suspicion, such as the Roma or people with Muslim-sounding names, who might be from the Caucasus.

Furthermore, the assumption of many normal Russian citizens is that they will stay near where they were born. A highly educated person living in, say, Tomsk is much more likely to spend his or her entire life there than would such a person born in Kansas City, Lyons, or Manchester, UK, be likely to stay in those cities. The idea of starting one's own company and freely seeking the place where the best facilities and personnel for that company might be found is still very weakly developed in Russia. Mobility of place and status are much more limited in Russia than in most other modern societies. This restriction in outlook and practice remains an obstacle to innovation.

14   The Legal System

The patent system added the fuel of interest to the fire of genius.
—Abraham Lincoln, February 11, 1859

The body of law on inventions, usually called patent law, is a turgid subject attracting relatively few historians or social analysts.[1] However, beneath the legal technicalities and obscure definitions is a subject of prime importance to modern countries, for it forces one to articulate important questions: how well does society encourage and reward innovation among its citizens, and how well does it protect inventors? Throughout its history, Russia has never adequately rewarded or protected its most innovative citizens.

Tsarist Russia was an autocracy. The tsar as autocrat (*samoderzhavets*) saw himself as the source of all power, whether wielded directly or indirectly. Therefore, "a privilege given to a merchant or inventor to receive income from a business or invention was not seen as a right, and certainly not as a property right, but as a dispensation, a charitable act, something given reluctantly and easily rescinded. The entrepreneur was never secure." The tsar and his close advisers feared the independent power that a truly successful entrepreneur or industrialist would inevitably possess. For this reason the tsarist government never recognized "patents," only "invention privileges." It never signed on to international patent conventions.

Despite tsarist Russia's failure ever to adopt an effective Western-style patent system, increasing concessions were given to inventors. In the seventeenth and eighteenth centuries most concessions given to people wishing to be active in the economy were to traders, who were sometimes awarded the exclusive right to run certain businesses and to trade in certain commodities. Then in the nineteenth century privileges were issued to people proposing new products or businesses; in 1812,[2] 1833,[3] 1870,[4] and 1895[5]

changes were made to the regulations governing invention privileges. All
these were rather minor modifications of the general principle that the
right to market a product was not allied with a property right possessed
by an individual but instead was a special favor bestowed by the government and serving the government's purposes. The government alone made
the decision as to whether the invention was worth pursuing. Of course,
all governments administer patent systems, hoping to foster innovation,
but the tsarist government was much less interested in promoting business
than it was in retaining its control and authority.

The process of getting such a privilege was exasperatingly slow and frustrating. An American inventor who applied for a Russian patent (actually,
an "invention privilege") was surprised years later to receive a response. He
congratulated the Russian authorities for their "good memory" and said he
had long ago forgotten about the whole thing.[6] As the leading scholar on
the subject of tsarist invention law observed, "The inventor was entirely at
the mercy of the bureaucratic machine."[7] As a result, few people received
invention privileges or even applied for them. In 1872 only seventy-four
invention privileges were issued in Russia, while 12,200 patents were
awarded in the United States.[8] The majority of such invention privileges,
often as high as 80 percent, were awarded to foreigners who wanted to sell
innovations in Russia. Foreign entrepreneurs were viewed suspiciously by
the Russian government. Several tsarist bureaucrats awarding the privileges
even justified the difficulty of obtaining them as a way of restricting foreign
influence in Russia.[9]

The length the privilege was in effect was shorter than in most other
countries, and the privilege could be revoked if it was not "worked," that
is, if the product was not actually marketed.[10] Since obtaining investor support for new industries was extremely difficult in Russia, many invention
privileges expired before any actual new product appeared. Thus, one of the
people featured earlier in this book, Pavel Yablochkov, obtained a privilege
for his electric light in 1878, in 1879 for new galvanic batteries, and in 1880
for a system of controlling electric current. But he was unable to find financial backing in Russia, and nothing commercially useful in Russia came out
of Yablochkov's outstanding innovative activity.

Pressure in the early nineteenth century to improve the invention privilege first came from the textile industry, especially spinning and weaving,
activities important to the Russian economy. In the late nineteenth century

a new organization—the Russian Technical Society, founded in 1866 and dominated by industrialists and engineers—began lobbying for more adequate protection of innovations similar to the protections offered by Western European countries, but the tsarist government continued to resist. As a result, there was mutual suspicion between the government bureaucrats and inventors. As one critic wrote, "The Russian inventor was faced with an almost insurmountable wall of ignorance and indifference, made worse by financial difficulties."[11]

Russian inventors were not lacking in brilliant and original ideas and theories, but developing the same into practical working inventions would often have entailed moving abroad. But if they moved abroad and then applied for invention privilege in Russia, they faced much discrimination. International conventions on patents were held in Vienna in 1873 and in Paris in 1883. Although the Russian government sent delegates, it refused to join any international conventions on the subject, not wishing to be bound by foreign obligations.

The secretary of the Scientific Committee of the Ministry of Finance, A. N. Gur'ev, openly advocated in the 1880s a policy of illicitly copying the best Western technologies under government protection, a policy that membership in international conventions on patents would have contravened. Nonetheless, foreigners continued to dominate in applications for invention privileges in Russia, and foreign equipment, especially German, and even foreign engineers continued to play very large roles in Russian industry to the end of the tsarist regime.

Soviet Russia was opposed to private property, and therefore the concept of "intellectual property" belonging to an individual was alien. The leader of the new Soviet state, V. I. Lenin, wrote sharp criticisms of Western patent systems, accusing them of helping capitalists exploit their employees and of slowing down the advance of technology through the use of "preventive" patents (a preventive patent is one that is taken out not to protect a new product but to protect an existing one by blocking production of a new competitor).[12]

On June 30, 1919, the new Soviet government issued what was called Lenin's Decree on Inventions, which declared all innovations to be the property of the Soviet state. A few years later, under the more permissive and temporary New Economic Policy, this decree was supplanted by a more lenient one, but that law was soon replaced by yet another, which

again banned ownership of intellectual property by Soviet citizens. Foreigners were still given some "patent" rights. In subsequent years (1930,
1931, 1941, 1951, 1959, 1961, 1966, 1968, and 1973), Soviet invention and
discovery law was modified, but the basic principles were now enshrined:
inventions were the property of the state; Soviet citizens could not have the
right to sell innovations or licenses for them.[13]

The Soviet government insisted that its system of invention law was
actually fairer than systems in the West, since individual innovators (as
opposed to companies) were rewarded with "author's certificates" designed
to honor inventors by recognition and even award them modest financial prizes if the invention was deemed particularly useful to the national
economy. However, Soviet inventors were given no period of monopoly
use and could not start a business. Furthermore, it was the government's
decision what was useful and deserved a prize. Governments everywhere
have proved to be poor judges of such things.

Since the goal of the Soviet system of invention law was to advance the
government's interests, the coverage of the law was much broader than
in most Western states. Author's certificates could be given not only for
inventions themselves but also for "technical improvements," "rationaliza-
tions," and "scientific discoveries." Thus, Soviet law on inventions created
what in most other countries would be called a "suggestion box" system:
any idea that an employee had for improving production or any advance in
knowledge was eligible for recognition. Of course, it was the management
of the state-owned company that made the decision about the value of
the suggestions, and disruptive suggestions were usually ignored. The most
valuable new technologies are disruptive, one reason why in the West they
often do not receive strong promotion within an existing company but
must be pursued outside by a new start-up.

The most interesting facet of the Soviet system was its coverage of "scientific discoveries," that is, fundamental advances in knowledge that may not,
at least at first, have had any practical application at all. The Western author
who has studied this aspect of Soviet law most closely, James Swanson, calls
it "one of those peculiar Russian institutions that make sense only within
the unique setting of a highly centralized socialist economy." He attributes
it to "a hyperpatriotic need to establish national priority in science."[14]

Because of the unusual characteristics of Soviet invention and discovery
law a few Western observers believed that it might indeed be "fairer" than

Western law, which often favors companies rather than employees. (We saw such favor in the case of Theodore Maiman and the laser.) One such observer and student of Soviet innovation law, Manfred Balz, concluded that this law "is not less just . . . in its treatment of creative individuals." However, Balz continued, he "was not concerned with the economic effectivity of the Soviet system."[15] That, unfortunately, is the rub. Seventy years of the Soviet system of invention law demonstrated amply that it was not economically effective. As the economist Joseph Berliner observed, the fatal flaw of the Soviet economic system was its inability to promote creativity and innovation.[16]

The frustrations of Soviet engineers who were so frequently unable to introduce their innovations into practice were perhaps most graphically illustrated in a famous 1956 novel by Vladimir Dudintsev titled *Not by Bread Alone*. In this novel an engineer who has developed a superior method of making metal pipes vainly tries to gain the attention of his employer or anybody else in the Soviet bureaucracy who might want to improve Soviet production methods. He finds out that Soviet administrators are primarily interested in production output, not improvements, because the administrators are rewarded for achieving increases in gross output. They oppose any innovation that would mean temporarily stopping the production line while new equipment is installed. Dudintsev's novel struck such a responsive chord among Soviet engineers (by that time the largest group of educated people in the Soviet Union) that it became a best-seller until repressed by Soviet authorities.

Only in the most recent years, after the collapse of the Soviet Union, has Russia moved toward a patent system similar to that in other industrialized countries. In 1992 Russia adopted a patent law, and in subsequent years a complex series of laws, edicts, and legal acts concerning intellectual property.[17] For the first time in the entire history of Russia, from Peter the Great to the present, it is now possible for Russian citizens to hold patents to their innovations in Russia—a patent, not an "author's certificate" or an "invention privilege." However, many problems remain because of lack of clarity, contradictions among the laws, and general inexperience with a patent system.

The law of 1992 was adopted when privatization of industry had not yet begun. Therefore a significant number of inventions of research institutes, industrial enterprises, and innovation firms that received rights to intellectual property on the basis of that law were, as earlier, property of the government. In that way, even though rights to intellectual property may have been transferred to the enterprise or research institute, the government remained directly or indirectly the owner of the property.[17] When privatization of many industries came later, it was very difficult to define the real owners of intellectual property that had been created earlier.

On military and dual-use technology the position of the government was much clearer. According to the legislation, rights to this type of intellectual property obtained at the expense of the government belonged entirely to the Russian Federation.[18] However, in 1998 and 1999 new legislation seemed to broaden the definition of what constitutes defense and national security in a troubling way. A special resolution of the government on September 2, 1999, stated that exclusive rights to any results of science and technology activity paid for by the government were assigned to the Russian Federation (the government) unless the exclusive rights had earlier been elsewhere assigned. An important difference could be noted between the legal acts of 1992–1993 on intellectual property and that of 1998–1999. In the earlier legislation, organizations had the title to intellectual property and the rights of the government were undefined. In the legislation of 1998–1999, the assertion of government rights to intellectual property was already the norm, the exceptions being those cases in which "rights had earlier been elsewhere assigned." What this meant in actuality was still unclear.

Since the majority of objects of intellectual property continued to be created with the financial help of the government, the administrators of most research organizations remained uncertain about their rights to it. A beginning to a clarification and normalization of the situation came at the end of 2001, when the government of the Russian Federation issued a new directive designating it as retaining rights to intellectual property created at government expense when those rights "are directly connected with defense and the national security of the government, and also those for which industrial application is a task assumed by the government"[19] (the exact meaning of the last phrase was ominously unclear). In all other instances the rights of the government to the results of R&D would be transferred either to the organization that developed the innovation or to another economic agent.

In recent years the Russian government has continued to try to create something resembling a Western-style patent system in Russia, and some real progress has been made. At the end of 2005 an important step was taken toward modernizing the legal regulation of intellectual property created by government funds. On November 17 of that year the Russian government adopted a resolution "Concerning the distribution of rights to the results of R&D." This resolution strengthened the rights of research organizations to the results of R&D performed at government expense.

An even more important step was taken on August 2, 2009, when Law 217 was passed, permitting research performed in educational and research institutions supported by the government to be commercialized, with rights to the intellectual property being retained by the institutions where the research was performed. This was an effort to duplicate in Russia the Bayh-Dole Act of 1980, the Patent and Trademark Law Amendments Act, in the United States, whereby universities can keep the intellectual property rights to the products of their research. Responding to this relatively new legislative act, many American universities have created technology transfer offices for the purpose of aiding the development of practical applications of research results, with subsequent commercialization.

The Russians in recent years have been very enthusiastic about the Bayh-Dole Act and technology transfer offices. In fact, in some cases Russian enthusiasm for this idea has exceeded that of the Americans who taught them about it. There are critics of the Bayh-Dole Act in the United States, people who fear the effects of commercialization on the educational function of universities, but this criticism gets little attention in Russia.[20] In addition, there is inadequate appreciation in Russia of the fact that in countries such as the United States, questions of intellectual property are still being hotly debated, especially in areas such as digital technology.[21] In Russia there is a tendency to emulate "finished systems" in other countries, not understanding that these foreign systems are not finished at all but are constantly evolving amid sharp controversies. And in their new enthusiasm for developing commercial applications of scientific research coming out of universities and other research institutions, following what they see as foreign practices, Russians often miss the strict rules that exist in most Western universities against abuses of these privileges. The possibility for corruption, already a major problem in Russia, therefore widens.

An example of this lack of rules was graphically illustrated to me in 2010 when I visited the computer science department of a leading Russian state university. There I learned that one of the professors was also running a successful private software company. When I asked where his company was located, he replied, "Right here, in the corridors of the university." In other words, this professor was using the facilities, services, and students of the state university for the purposes of his private profit-making company. When I asked if he did not feel that he owed the university something for the use of its electricity, offices, computers, and talent to run a private company, he replied, "I owe the university nothing. They should be glad to have me." An American lawyer accompanying me was aghast, and said, "Basically, there are no rules." Although my home university, MIT, is an avid commercializer of research results, such an abuse would never be permitted there.

Thus, although Russia has made great progress in the last twenty years in developing laws on patents, intellectual property, and the commercialization of technology, the legal system is still weak and poorly understood, the patent laws are still untested, and there are even unresolved contradictions between them and the rest of the legal system. It is easy for everyone—businesspeople, university professors, government officials—to abuse the rules since no one knows quite what they are. Entrepreneurs are still insecure, fearing that if they gain too much power and influence the state will move against them, as it did by arbitrarily imprisoning (not once but twice) the man who was once post-Soviet Russia's richest and most able businessman, Mikhail Khodorkovsky. Russia still has a long way to go before it achieves an adequate legal system covering innovation and business.

We see that the influence of the legal system on innovation is much larger than merely patent law. Businesspeople need to feel protected by the law in all their operations, not just their use of innovations. People who are accused of a crime need to feel they have a chance of being acquitted. The fact that the whole legal system in Russia is subject to political influence and that judges are not truly independent raises questions much larger than laws on patents and intellectual property.

15 Economic Factors

The OECD economies are increasingly based on knowledge and information. Knowledge is now recognized as the driver of productivity and economic growth.
—"The Knowledge-Based Economy," Organisation for Economic Co-operation and Development[1]

The lack of an adequate patent law system throughout Russia's history has been such an obvious obstacle to commercialization of technology that some people may conclude it is the most important of all problems. One recent popular author called the patent system "the most powerful idea in the world," describing it as a system that in Abraham Lincoln's words "added the fuel of interest to the fire of genius."[2] However, this elevation of the patent system to the status of the single greatest spur to innovation is an exaggeration. The legal protection of inventions is only one of many impetuses needed for technological progress. Attitudinal, social, economic, and political obstacles are just as significant in Russia as is the absence of an effective patent system. An inventor with a patent who cannot find investors is destined for failure; the patent cannot do the job by itself. In the final analysis, the reason why innovators like Cherepanov, Yablochkov, Sikorsky, and Losev were not able to succeed in Russia was not the absence of a patent system; it was the lack of interest in and support by investors. And behind this missing financial support lay many other factors, including attitudinal, social, and political ones. Tsarist Russia did not have an investment class, Soviet Russia banned the existence of such a class, and post-Soviet Russia has a very weak one.

What kind of an economic system spurs innovation? In the last two generations a new emphasis has developed in Western economic thought on "innovation economics," a trend of thought that can be contrasted to the classical economics of Adam Smith and the "mixed" economics of John Maynard Keynes. Innovation economics places technological change and its resultant increases in productivity at the heart of economic growth. Capital accumulation, so often stressed by economists in the past as the key to growth, is seen as less important than innovation. We now talk about "knowledge-based economies."

In Russia today there are few strong stimuli for investing in high-technology projects, either for the government or for the few private individuals who might do so, at a time when it is significantly safer to invest in resource-extracting industries such as oil, gas, and minerals. These extractive industries are the genuine strengths of the current Russian economy. (Russia now produces as much oil as Saudi Arabia, and in some time periods is the top producer in the world.) Thus, the strength of extractive industries depresses the position of innovation industries, soaking up much of what capital is available. Some people call this the "oil curse." To overcome this obstacle would require a major change in economic incentives, which so far the Russian government shows little sign of implementing.

Compared to other industrialized countries the share of business financing of R&D in Russia is low and the share of government financing of R&D is high. Furthermore, in recent years this disparity has increased, not decreased. In 2005 gross domestic expenditures by Russian businesses on R&D was 22.4 percent, while the government share of such R&D was 60.1 percent. Five years later, in 2010, the business share was 18.3 percent, while the government share was 68.8 percent.[3] (By comparison, in 2010 in the United States the government share was 27.1 percent and the business share was 67.3 percent.) Thus, at the moment the hopes that Russia will become a high-tech country based on business investment are diminishing, not growing, at least as seen in these economic statistics.

Other obstacles in Russia to venture capital investing are the lack of experienced managers of venture capital; an undeveloped securities market, the short time horizon of those investors who do exist, weak protection of intellectual property, the unwillingness of many innovators (scientists, engineers) to give up controlling interests in their innovations, and the red tape and legal obstacles involved in complying with government regulations.

As a result of all these obstacles, those few venture capitalists who are active in Russia usually choose their investments on the basis of family connections or friendships rather than through an analytical approach to all possibilities existing at any given time. Furthermore, the unreliability of available statistics often makes determining the profitability of past investments difficult. Not surprisingly, few large investments are directed toward new companies, finding their way instead to already existing ones in second or third rounds.

Nonetheless, Russia is making a major effort at the present time to soften or eliminate these obstacles to investment in innovation, and the government has created several new institutions and agencies directed at helping create a "knowledge-based economy." These efforts will be discussed in the last section of the book, "Can Russia Overcome Its Problem?"

16 Corruption and Crime

"Most Russians have grown so accustomed to a certain lawless way of life that they come to view corruption as Russia's own special way."
—Misha Friedman, *New York Times*, August 18, 2012

Russia is among the most corrupt nations in the world. According to the Corruption Perceptions Index published by Transparency International, in 2011 Russia placed 143rd out of 180 countries.[1] The practice of paying off officials is not new; in tsarist times corruption was even legal at many moments, since bribes paid to local officials were often seen as their major source of income. There was even a word for this way of maintaining a bureaucracy: *kormleniia*, or "feedings." It is true that certain rulers, such as Peter the Great and Catherine the Great, tried to combat the practice, but until the end of the empire corruption was widespread.

The Soviet government tried to eliminate corruption by making bribe-taking a capital crime, but the practice of bribes was still common. In post-Soviet times corruption has spread widely. According to some polls, about half of the adult population of Russia admits to have given bribes. Protection money became a common obligation for the new private businesses that sprang up in Russian cities and towns after the fall of the Soviet Union. Gangsters and sometimes the police themselves would warn merchants that unless they paid for protection, some unforeseen disaster might strike their stores, such as smashed windows and stolen goods. Most people took the safe way out. An acquaintance of mine, Paul Tatum, the American manager of a Radisson hotel in Moscow, refused to pay protection money; he was gunned down in the street a block from his hotel. I have had the experience of riding in a car with a friend in Moscow when a policeman flagged us down for some unstipulated infraction. My friend placed his driver's license and several high-value bills on the dashboard for the policeman to choose as he wished. After checking the documents and taking payment the militiaman waved us on.

Corruption is not only a drain on the Russian economy, it directly affects its ability to compete in high technology, blunting the excellence that is needed to survive in world competition. Getting licenses for launching entrepreneurial ventures requires permission from judicial bodies, which often requires bribes. Contracts are not usually awarded without a kickback (*otkat*). Entrance into higher educational institutions and the granting of diplomas are frequently facilitated through bribes. We saw a dramatic example of such corruption of technical qualifications in the chapter on aviation; in 2010 Russian newspapers and the *New York Times* reported that seventy of the engineers at the famous Sukhoi aviation company bribed educational authorities at a technical college to give them false engineering diplomas. The reputation of Sukhoi, striving to enter the international market with a new passenger jet, was severely damaged.

Tax inspectors and regulatory agencies are notoriously arbitrary, modifying their behavior in response to the "rewards" they receive. All this means that instead of the best businesses and entrepreneurs prospering, often it is merely those able to pay, sometimes because they are engaged in criminal activities themselves.

Although crime is not as rampant in Russia today as it was in the early 1990s, it is still widespread. Common crimes such as theft and homicide are problems in many countries, including both Russia and the United States. However, in Russia, crime and corruption affect entrepreneurship, a subject of this book, more than in many other nations. Any successful entrepreneur becomes a likely target for criminal groups. We saw in chapter 10 that Eugene Kaspersky, perhaps the leading software entrepreneur in Russia today, was targeted in 2011 by the kidnapping of his son and the demand of a ransom. Kaspersky outsmarted his enemies, but this was a rare example of such success.

Charges of corruption have already been leveled against one of Russia's newest efforts to create high technology, the Skolkovo project, to be discussed in chapter 19. In February 2013 the finance director of the Skolkovo Foundation, Kirill Lugovtsev, and the head of the Skolkovo Customs Finance Foundation, Vladimir Khokhlov, were accused of embezzlement.[2] And the Russian head of the Skolkovo Foundation, the oligarch Victor

Vekselberg, was accused of depositing Skolkovo funds in his own bank for his own benefit.[3]

Criminality and corruption of a special sort penetrates the Russian government itself in a way that inhibits independent activity by Russian citizens. The tax and economic crime laws are imposed arbitrarily and selectively. A favored tactic of the Putin government in suppressing troublemakers and critics is to accuse them of economic crimes, illustrated most dramatically by the imprisonment of Mikhail Khodorkovsky, but there are many other examples. Knowing how easy it is to fall afoul of these authorities, an engineer or a scientist with an idea for an innovation may be reluctant to enter business. To do so is often seen as providing authorities with opportunities for selective prosecution if the new entrepreneur does anything the government does not like. The damping effect on innovation is obvious.

17 The Organization of Education and Research

If the eighteenth century was the century of academies, while the nineteenth century was the century of universities, then the twentieth century is becoming the century of research institutes.
—S. F. Ol'denburg, permanent secretary of the Soviet Academy of Sciences, in a 1927 report to the Soviet government[1]

What is the optimal way of organizing research and development in order to achieve industrial innovation? Although no one knows the answer to this question with certainty, I argue that Russia has not been in step with world trends in organizing knowledge expansion and consequent technological progress, and it has paid a heavy price for this mistake. Misled by some European developments at the beginning of the twentieth century, it created a system that is strong in promoting theoretical science but very weak in translating that knowledge into industrial applications. The organizational causes of this weakness are still poorly understood in Russia, and often not very well appreciated outside that country. To understand these causes more fully we must look briefly at world trends in the promotion of technological progress. We will see that some of the problems that afflict Russia today also are present in other countries, and therefore an alleviation of these problems has general significance.[2]

In the eighteenth century, "academies" were often seen as the best loci for the work of a country's leading scientists. In the nineteenth century, starting in Germany and then elsewhere, research universities became more and more important and gradually pushed aside the academies as the actual place of research in most countries, although Russia lagged in this development. In the twentieth century the idea began to grow in industrialized countries that teaching in the universities constituted a load on talented scientists that prevented them from devoting themselves properly to research, and therefore the concept of the nonteaching research institute gained prominence. In France, one of the early models of this new type of organization was the Pasteur Institute (1887); in Germany, an outstanding example was the Koch Institute (1891). In the United States, emerging private foundations and wealthy individuals helped answer the call of leading scientists for new preserves for unfettered research. Leading science administrators did not realize at that time that giving the scientists what they called for might not be best for the welfare of the nation as a whole.

In the United States, John D. Rockefeller in 1901 backed the creation of the Rockefeller Institute, in which he said he wanted scientists "to work in utter freedom." There would be no bothersome students, no teaching loads, few committee meetings, no outside obligations. Several other institutes followed in the United States, including the Carnegie Institute in Washington, DC (established one year after the Rockefeller Institute, in 1902), and, quite a bit later, in 1930, the Institute for Advanced Study in Princeton. These institutions were modeled on the new European research institutes, particularly the German and French ones. In the case of the Rockefeller Institute, the Koch and Pasteur institutes were specifically cited by the original Rockefeller trustees as they created their new research organizations.[3] In Germany, institutes as a model for research received an important new impetus with the creation in 1911 of a whole network of them in the Kaiser-Wilhelm Gesellschaft, the predecessor of the present-day Max-Planck Gesellschaft.[4]

In the new Soviet Union of the 1920s the idea of planned research conducted by the government was especially prominent, and the new nonteaching research institutes seemed to be the way to do this. After returning in 1926 from an inspection tour of scientific institutions in Germany, France, and England, S. F. Ol'denburg, permanent secretary of the Soviet Academy of Sciences, wrote a report for the government in which he said, "If the eighteenth century was the century of academies, while the nineteenth was the century of universities, then the twentieth century is becoming the century of research institutes."[5] The Soviet Academy of Sciences, containing at first dozens and then later hundreds of such institutes, was planned to be different from the learned societies of the eighteenth century, and also different from the universities of the nineteenth century; instead it would be a sort of "ministry of science" based on research institutes in which scientists did not have teaching duties but would instead devote themselves to the advancement of knowledge.

For a while it looked like the creation of such networks of nonteaching institutes was a worldwide trend. In France in 1936 a left-wing government came to power, dominated by socialists and communists. This government was very impressed by the new science establishment of the Soviet Union, and it created a scientific organization in France partly modeled on it that still exists today, the Centre National de la Recherche Scientifique (CNRS), which is a network of government-supported nonteaching institutes.[6]

However, as time went on, the allure in the West of the nonteaching institutes began to wane, especially in the United States. The feeling began to spread among academic administrators that teaching, far from being an impediment to research, was actually a stimulus to it. Senior researchers who did not have to meet the challenges to their ideas that students sometimes bring often settled into intellectual ruts in which they pursued the same idea endlessly. Nonteaching institutes in the United States, such as the Rockefeller Institute in New York City, the Institute for Advanced Study in Princeton, or the Carnegie Institute in Washington, DC, became more and more exceptions to the general pattern. They definitely did not fulfill the grand dreams of some of their originators of becoming the new model for research throughout the country.

In the United States the research universities did not decline, as Soviet planners had predicted, but instead developed into the most powerful engines of knowledge the world has ever seen. The infusion of federal money that came during and after World II played an enormous role here. Universities and the federal government in the United States developed a remarkably successful complementarity of functions in the promotion of science.[7] The competitive peer-review process required in granting much of the federal money awarded to university researchers provided a mechanism for ensuring quality research. It is difficult for the director of an institute to diminish the state-provided budget of a veteran member of his or her research staff, even if that scientist is obviously not doing as well as previously; it is easy for the National Science Foundation to turn down the proposal of a university researcher if the peer-review panel submits critical assessments. The system of funding research forming in the United States after World War II was simply more effective than the older systems, both at home and abroad.

A symbolic indication of shifting attitudes toward the relative merits of nonteaching research institutes and research universities came in 1953 when the board of trustees decided to convert the nonteaching Rockefeller Institute in New York City into a university with students.[8] The trustees at that time conducted an examination of the achievements of the Rockefeller Institute and concluded that it was not meeting its full potential. It is interesting to look at the comments that were made in the reports to the Rockefeller board at that time and compare them with recent criticisms that have been made of the institutes of the Russian Academy of Sciences, which were (and are) primarily nonteaching research establishments (the Russian Academy of Sciences remains Russia's premier research institution)[9]:

Research groups in the Institute were "cloistered" and "isolated."
Too much inbreeding of ideas and people.
The atmosphere in the research labs does not have the freshness that the youthful enthusiasm of students can bring.
Students have zany ideas, but out of every 100 such ideas, one or two turn out to be fundamentally important.

As a result of their reassessment the Rockefeller board decided to convert the Rockefeller Institute into Rockefeller University, complete with students. Quite a few of the existing researchers disagreed with this decision and refused to take on students, but within five years it became clear that they were losing out in winning grants to their faculty colleagues who had students in their labs. This was a revelation, as the nonteaching senior researchers at first considered their teaching colleagues to be less distinguished intellectually. But these teachers were more successful in getting grants, they were creating more exciting laboratories, and they were publishing more valuable results. Once this difference became clear, the resisters capitulated, and the conversion of the Rockefeller Institute to Rockefeller University was complete.[10]

In the United States, the preference for research universities over nonteaching research institutes as the locus for both teaching and research has in recent decades become more and more clear. When Richard Feynman, a world-famous physicist, was offered a nonteaching position at the Institute for Advanced Study in Princeton, he turned it down. Feynman observed,

I could see what happened to those great minds at the Institute for Advanced Study, who had been selected for their tremendous brains and were now given this opportunity to sit in this lovely house by the woods there, with no classes to teach, with no obligations whatsoever. Those poor bastards could now sit and think clearly all by themselves, OK? . . . And nothing happens. Still no ideas come.

Nothing happens because there's no real activity and challenge. . . . Students are often the source of new research. . . . Teaching and students keep life going, and I would never accept any position in which somebody has invented a happy situation for me where I don't have to teach. Never.[11]

Another example of this preference can be found in the views of Roald Hoffmann, Nobel laureate and professor of chemistry at Cornell University (and a foreign member of the Russian Academy of Sciences), who in 1996 wrote, on the subject of teaching and research,

not only are the two inseparable, but teaching makes for better research. . . . I am certain that I have become a better investigator, a better theoretical chemist, because I teach undergraduates.[12]

This personal and anecdotal support of the merits of combining teaching and research received a more rigorous confirmation in 2011 when the prestigious journal *Science* published an article comparing two groups of graduate students: those with both teaching and research responsibilities and those with only research responsibilities. The authors found that teaching can contribute substantially to the improvement of research skills. The graduate students who did both teaching and research demonstrated "significantly greater improvement in their abilities to generate testable hypotheses and design valid experiments."[13]

The full significance of these international developments in research organization has not yet dawned on Russia. In Russia, nonteaching research institutes of the Academy of Sciences still enjoy more prestige and privileges than the state universities, which traditionally have been primarily teaching, not research, institutions. Great "research universities" do not exist in Russia. Even Moscow University and St. Petersburg University, generally considered the best universities in the country, rank low in international rankings of research productivity.[14] Thus, just as Russia is lamed by an industrial system inherited from Soviet times that is inefficient, so also it is hobbled by a scientific research establishment that is not in line with best practices abroad and that does not produce as much bang for the ruble as it should. Research institutes in the Russian Academy of Sciences are typically large institutions with low "impact factor" ratings in the world scientific literature.

The analysis of Russian technology in this book has had the unexpected
· effect of highlighting some of the characteristics of technology in the West,
often compared here to Russian technology. In Russia, societal constraints
(attitudinal, political, social, economic, legal, and organizational) have
often caused promising technological starts to falter, to fail to be developed
further. In the West, a different society has had other effects, not all good.
On the one hand, the Western emphasis on competition, patents, entrepreneurship, and economic success has driven technology forward. On the
other hand, these same factors have caused endless fights over priority,
lengthy and very expensive litigation over patents, and bitterness among
researchers, in some cases people who had once been close friends.

In the development of electricity in the United States, George Westinghouse and Thomas Edison engaged in a sharp conflict, "the war of the
currents." Edison spread disinformation about Westinghouse's alternating
current system, and described electrocution as being "Westinghoused."[15]

In the invention of the transistor in the United States the three men
usually credited with this achievement—William Shockley, Walter Houser
Brattain, and John Bardeen—soon broke apart because Shockley claimed
the lion's share of the credit. When the men met in Stockholm to jointly
receive the Nobel Prize, it was the first time in a long while that all three
had talked together.

In the development of the laser, the researchers most often credited
with important achievements—Charles Townes, Arthur Schawlow, Theodore Maiman, and Gordon Gould—fell into similar disputes. Townes,
Maiman, and Gould each claimed priority for himself and subtracted from
the achievements of the others. Maiman wrote:

In real-world science, intense competition for recognition, credits, and budgets
abounds. Perhaps not surprisingly, the reactions from unsuccessful competitors often come out more like political "spin" than science, dirty tricks and all. Intrigue in
science may not be what most of us expect, but such is the reality.[16]

Of course, intrigue existed in Russia as well. But since the reward system there was dramatically different, the intrigue went along different
lines. Because personal fortunes were not to be made from technology, the
rewards most sought were institutional status and political connections.
The two Russian Nobelists in the field of lasers, Alexander Prokhorov and
Nikolai Basov, after their major achievement found that they could not
exercise their egos adequately in the institute of the Academy of Sciences

where they had originally worked together, the Lebedev Physics Institute, so eventually each was made director of his own institute. There each could have the full institutional authority he sought.

In the post-Soviet period the major figures in industry often also occupied positions in government as well, so they had both political and economic power. They often had offices in the Kremlin. Maintaining favored political positions was a paramount goal for industrialists, as Russia's richest man, Mikhail Khordokovsky, found out to his regret, ending up in prison for years. Different societies have different reward systems, and the results display both strengths and weaknesses.

III Can Russia Overcome Its Problem Today? Russia's Unique Opportunity

Russia today has the greatest chance in its history to break out of its centuries-old pattern of technical brilliance followed by commercial failure. Already the ideas of its scientists and engineers are "less lonely" because of multiplying connections between Russian researchers and Western companies seeking technological innovations. But who will be the primary beneficiaries of these new links, Russia or Western companies and investors? Although in a globalized world the concept of "national" companies is much less clear than it once was, Russia has a long way to go to win its place either among the giant international companies that now bestride the globe or as a birthplace of exciting startups.

Meanwhile, the Russian government has set an almost impossible goal: converting Russia in a short period of time from a country primarily dependent on natural resources for economic prosperity to one relying on a knowledge economy.[1] Such a transition is enormously difficult. Let us look at some of the steps that Russia is taking toward that goal.

18 Creating New Foundations and Research Universities

The most important activity of the foundation is promoting long-term stipends and grant programs for talented students and promising teachers of the leading state universities of Russia.
—Website of the Russian Potanin Foundation (http://www.fondpotanin.ru)

New Foundations

The Russian government, observing what it considered to be the success of research in the United States supported by entities such as the National Science Foundation, broke with the exclusive Soviet tradition of block funding of scientific institutions from above and created new science and technology foundations using the grant system. These new institutions relied on applications from individuals or groups of scientists who wished to pursue research on specific problems, a departure from the old Soviet tradition of central government direction of science. It is difficult for Westerners, long accustomed to peer-reviewed grant systems, to realize what such a change meant, at least in principle: suddenly, each researcher or research team became free to submit an application independently and could solicit government funding for a research project designed by local scientists. The problem has been, however, that the amount of money channeled to scientists and engineers in this way has been limited; much of Russian research is still centrally funded and directed.

Among the new foundations were the Russian Foundation for Basic Research, the Russian Foundation for the Humanities, the Russian Fund for Technological Development, the Fund for Assistance to Small Innovative Enterprises, and the Venture Innovation Fund. A number of Western firms working with these new Russian foundations have found niches where cooperation is very fruitful. For example, Maxwell Biotech, a Russian venture fund, is helping American biotechnology firms conduct clinical trials in Russia of new drugs designed to combat cancer, liver disease, and hepatitis. In return for clinical trial opportunities, favorable licensing arrangements can sometimes be made between American and Russian companies. As the Russian managing director of Maxwell Biotech told an American audience on March 13, 2012, "Because of the regulatory environment in the United States, clinical trials can sometimes be conducted more rapidly and efficiently in Russia than in the United States."[1]

A Russian colleague and I have written fuller descriptions of these new Russian investment organizations elsewhere, and an ample literature on them is available.[2] Despite worries about regulatory laxness, these funding agencies are definitely steps in the right direction, but they are plagued by problems, especially low funding and favoritism in choosing recipients. The peer-review system has been introduced to Russia, but it still works imperfectly; it is common practice, for example, for applicants to write their own letters of reference and submit them to reviewers, who are supposedly making objective evaluations.

Even a few nongovernmental foundations have emerged in Russia, such as the Dynasty Foundation and the Potanin Foundation. Private philanthropy is new to Russia, but it is emerging. However, private philanthropy has very limited influence in science and technology. Foreign foundations have also been active in Russia, giving some support to scientific research.

In summary, foundations have had some good, but still inadequate, effects on stimulating initiatives by individual researchers to embark on promising research leading to new technologies.

Research Universities

Russia does not have a true research university system. The best researchers are not in the universities but in the Russian Academy of Sciences, which is more prestigious than the university system. To be an "akademik," a full member of the Academy, has traditionally been one of the greatest honors an intellectual can receive. Although the Academy has many critics, especially because of its authoritarianism, conservatism, and inability to commercialize technology effectively, it has fought hard to maintain its status, usually successfully.

One of the most interesting reform attempts in Russian science in recent years has been the effort to strengthen research in Russian universities, particularly in comparison to the Russian Academy. This effort is one in which I have been intimately involved, working together in a multi-million-dollar program with the John D. and Catherine T. MacArthur Foundation, the Civilian Research and Development Foundation, The Carnegie Corporation of New York, and the Russian Ministry of Education and Science.[3] In the process I have spent time at numerous Russian universities, all over the country. The program was at first jointly financed by Americans and Russians. When the Americans first suggested this program over fifteen years ago, we were told by our Russian colleagues that we should not use the term "research university," a comment that befuddled us, since the creation of such universities was what we thought the program was all about. The Russians told us that such a term was seen as too controversial and even threatening in Russia because the Academy of Sciences believed it had a monopoly on good research and wanted the universities to be primarily pedagogical institutions whose most talented graduates would end up in the Academy. One of the signs of progress of this program is that the Russian authorities have abandoned this position and currently have a program of their own designed to create research universities.

When we began the program,[4] the first problem we had was how to create research centers in the universities that could compete successfully with local Academy of Science institutes (of which there are hundreds). We did not have nearly enough money to build up the total budgets of the several dozen major universities in Russia. We chose another, more selective path. We looked at the universities, and in each one of several (later sixteen, then twenty) we chose a department that had some promising, well-qualified faculty members.

The first university we chose was Nizhny Novgorod State University, which had a good physics department. We (that is, the program, known as BRHE, for Basic Research and Higher Education) gave that department a piece of expensive equipment that the local Academy institutes did not have, the latest version of a scanning probe microscope, necessary for conducting research in nanotechnology. Suddenly the prestige roles of the university and the local Academy institutes were reversed; now researchers from the Academy came knocking at the door of the university asking if they could "get a little time" working with this equipment. We then launched a research program at the university using the equipment in which local Academy institutes were invited to be involved, but with the university playing the lead role.[5] And we helped establish technology transfer offices at these universities, a novelty for the Russians, that would look for ways to commercialize the research results.

This pattern was then repeated at other Russian universities, and the program has been so successful that eventually the Russian Ministry of Education and Science offered to take it over completely, and the Americans gradually receded as funders. Today the program is almost entirely funded by the Russians. Successful as the program is, however, by itself it has not been able to accomplish the enormous task of creating a research university system in Russia. After all, the program affected just a few departments and ran counter to the prestige-ranking tradition of Russia and the reputations of powerful individuals, especially those in the Academy. Nonetheless, a start has been made. It is still far too early to tell how the program will work out. If Russia could energize its universities, where talent first manifests itself, it could greatly improve its capabilities in entrepreneurial technology. When I visited St. Petersburg State Polytechnical University in December 2011, I met a few undergraduate students who spoke of their desire to become entrepreneurs.

19 RUSNANO (Nanotechnology) and Skolkovo (a New Technology City)

Russians still cannot grasp the idea of making money simply by inventing something. The mission of Skolkovo is to show people that they should not be afraid of starting their own business. Risk shouldn't be confounded with danger. There's always a risk; danger is very often just perceived.

—Pekka Viljakainen, adviser to Viktor Vekselberg, head of the new technology city of Skolkovo, speaking in 2012[1]

In recent years the Russian government has launched several programs aimed specifically at high technology, the largest and best known of which are RUSNANO, an effort to capture the promise of nanotechnology, and Skolkovo, an effort to create in a newly constructed technology city a Russian version of Silicon Valley.

RUSNANO

A significant change in technology worldwide has come from a new ability to manipulate matter on the molecular scale, or nanotechnology. The instruments that make this possible include scanning tunneling microscopes and atomic force microscopes, along with other rapidly developing methods. In recent years there has been a great deal of publicity about nanotechnology.[2] Some people say that nanotechnology will usher in another Industrial Revolution, improving old products and developing entirely new ones. Whatever the truth, most knowledgeable observers agree that nanotechnology is a development of enormous significance for all industrialized nations. In the United States, the National Nanotechnology Initiative (NNI) was created in 2001 to coordinate federal nanotechnology research and development. Between 2003 and 2010 the NNI invested $12 billion in nanotechnology projects, an enormous amount of money. The NNI has been described as "the single largest federally-funded multiagency scientific research initiative since the space program in the 1960s."[3] So we are talking about BIG science and technology research when we talk about nanotechnology.

Dmitry Medvedev, president of Russia during the peak of interest in nanotechnology, placed great hopes on Russia's use of nanotechnology to modernize the Russian economy and decrease its dependence on oil. In 2009 at a conference on the subject in Moscow attended by over 11,000 people from thirty-eight countries he said,

Nanotechnology will rival oil as a global powerhouse industry, so Russia's economy needs to embrace it now. . . . The global nanotechnology market is worth about $250 billion today and may reach $2 trillion by 2015, making it comparable to the market of natural resources. . . . We here [in Russia] have the knowledge, the financial resources and the administrative capacity to become leaders in a technological process that will change the world.[4]

To promote nanotechnology in Russia, the Russian government created in 2007[5] a special organization called RUSNANO, which is somewhat similar in its goals to the NNI in the United States, created in 2001. While the two organizations have a similar goal—namely, to make their respective countries leaders in nanotechnology—the two countries pursue that goal in rather different ways. Let us examine briefly RUSNANO.

The first thing that one should notice about RUSNANO is that it has been given a great deal of money, illustrating how important the Russian government considers the initiative to be. That sum as of 2012 was about U.S. $18 billion, which would make governmental investments in nanotechnology in Russia more than that of Japan or China, which also have big nanotechnology programs, and near that of the United States and Europe. Vladimir Putin, the most important political leader in Russia today, has said, "Nanotechnology is an activity for which this government will not spare money."[6] Of course, government investment is not the whole story, as private investment is also very important in the field, and this is an area where Russia definitely lags.

The man chosen to lead RUSNANO was Anatoly Chubais, a very prominent and controversial man. In the Yeltsin administration Chubais was in charge of the privatization of industry and devised the voucher privatization scheme that resulted in the creation of the Russian oligarchs, a few very rich men who managed to get control of much of the Russian economy. For his role in this scheme Chubais was roundly hated by many people in Russia, and in fact was the victim of an unsuccessful assassination attempt in 2005. Despite the controversy surrounding his name, Chubais is respected as a gifted manager, and since September 2008 he has been a member of the advisory council for JP Morgan-Chase. Before that time he was the head of the state electrical energy monopoly.

Chubais was the architect of RUSNANO's scheme to modernize Russian technology through the broad introduction of nanotechnology. Wanting to learn American methods of management, he came with RUSNANO's top administrators to study briefly at MIT's Sloan School of Management (where I was one of the MIT faculty who lectured and talked to them).

The way the RUSNANO program works is as follows. RUSNANO calls itself a "corporation," but it is not an operating company but instead acts as a foundation funding nanotechnology ventures. RUSNANO can invest up to 50 percent minus one share in a start-up's capital. Its goal is not to maximize profit but instead to set up other companies, and then to exit from them as soon as they can stand alone. RUSNANO makes money available not only to Russian citizens and institutions but also to citizens and institutions in other countries, including the United States (it has, for example, agreements with Alcoa and the Dow Chemical Company). Originally, there were two requirements for anyone wishing to get money from RUSNANO. First, the project had to involve nanotechnology, and second, the project had to involve a production facility in Russia, in other words, a factory. This last requirement was crucial, and it reveals a major goal of RUSNANO: not just to invest in research (although it can do that) but mainly to create a nanotechnology industry in Russia.

In recent years RUSNANO has relaxed its requirements just a bit. Now any company getting money from RUSNANO must establish either a production facility or an R&D facility in Russia. A further relaxation came in 2010 when its board allowed RUSNANO to invest overseas in order to develop global markets. It has funded twelve overseas investments, for a total of $2.7 billion, and it has funded five venture funds in the United States for $1.8 billion. Despite this overseas activity, the main emphasis of RUSNANO remains developing high-technology production facilities in Russia.

RUSNANO as of March 2012 had received almost 2,000 applications, including 372 from 37 countries, with 135 coming from the United States. Of these, 140 projects were approved, most Russian, with a total investment of $18 billion.[7]

The composition of the selection committee, called the Scientific-Technical Council of RUSNANO, was obviously important. The nineteen members of this council were some of the most distinguished research administrators in Russia; about half of them were members of the Russian Academy of Sciences; the rest were drawn from universities and industrial and defense research institutions.

In its first years of operation RUSNANO approved a number of projects, including:

Flexible polymer packaging (using nanocomposites)

Nano-ink production

Cutting tools using cubic boron nitride nanopowder

Solar batteries

Nanovaccines

Gallium-arsenide substrates

Very large-scale integrated (VLSI) circuits

Abrasive resistant parts using nanostructured ceramics

Radio-frequency identification (RFIDs) tags

Bifacial monocrystalline solar modules

Drugs for oncology therapy

Catalytic production of fuel from carbon dioxide without biomass

Energy-efficient high-voltage composite power lines

90-nanometer microchips

Film wrap for food preservation

Superstrength springs

Vertical lasers for high-speed fiber optics production

I will make some generalizations about the Russian nanotechnology effort, but before I do so let me turn to the second major Russian effort to energize the country in high technology, the Skolkovo project.

Skolkovo: The New Technology City

Skolkovo is often described as "Russia's Silicon Valley." It is a large area of several thousand acres in the suburbs of Moscow until very recently devoted to agriculture. The Russian government is investing several billion dollars here to build an "innovation city" (*innograd*) at the center of which will be a new university. The Skolkovo Foundation, which is setting up and running the whole project, is chaired by two people, one a Russian and one an American. The Russian chair is Victor Vekselberg, one of the Russian oligarchs, a multibillionaire who made his fortune primarily in oil and aluminum but who has many other assets as owner of the Renova Group of development companies. The American chair is Craig Barrett, former chair of the board of Intel and before that its fourth president. Skolkovo also has a Scientific Advisory Council, and that too is jointly chaired by a Russian and an American, both of them Nobel Prize winners. The Russian is Zhores I. Alferov, who won the 2000 Nobel Prize in Physics (which he shared with Herbert Kroemer and Jack S. Kilby) and is credited with inventing the heterotransistor. (For an illustration of Alferov's retrograde political views, see chapter 6, on transistors.) The American chair of the Scientific Advisory Committee is Roger D. Kornberg, a Stanford professor who won the Nobel Prize in Chemistry in 2006 for his studies of eukaryotic transcription.

The fact that both the foundation and the advisory committee are jointly chaired by Russians and Americans demonstrates that Skolkovo is an unusual effort by the Russians to be international. As Vyacheslav Surkov, a top adviser to the primary proponent of Skolkovo, Prime Minister Dmitry Medvedev, said, the new innovation city will be designed to "import the best brains from all over the world" and will provide luxurious accommodations and cultural attractions that will make it a desirable destination. Russian newspapers have run dozens of articles describing Skolkovo as a futuristic place, with modern buildings, unusual transportation schemes, research labs, a university, and high-tech companies.

The goal of Skolkovo is to be not only a leading research center but also a place where scientific advances are quickly commercialized and introduced onto the market. With that goal in mind, the Russian government has promised foreign companies that if they set up branches in Skolkovo, they will be freed from the normal taxes and given a variety of additional incentives. Responding to these invitations, many foreign companies have accepted, including Boeing, Intel, Siemens, Nokia, Samsung, and Cisco. Cisco has promised to invest $1 billion in its facilities in Skolkovo. Many countries are now involved in cooperation with Skolkovo; during his state visit to China in June 2012, Vladimir Putin signed such an agreement with the Beijing-based Z-park.[8]

At the center of Skolkovo is a new university, one that is expected to play the role of Stanford in Silicon Valley or MIT in the Boston area. After visiting both Stanford and MIT, the administrators of Skolkovo chose MIT as their partner. They particularly liked MIT's success in spawning start-up companies, usually about twenty or thirty a year. In 2010 and again in June 2011, MIT and the Skolkovo Foundation signed "preliminary agreements" for cooperation and worked toward a definitive, legally binding agreement. In October 2011 MIT's president, Susan Hockfield, signed in Moscow a three-year collaboration agreement with Skolkovo.

The Skolkovo Institute of Science and Technology (SkolTech), on which MIT is cooperating with the Skolkovo Foundation, is a graduate-level university combining education, advanced research, and entrepreneurship. Although the university itself will be located in Skolkovo, near Moscow, it will work closely with other universities in Russia and abroad. The educational program will be organized around five broad themes rather than traditional academic disciplines. Those themes are (1) energy science and technology, (2) biomedical science and technology, (3) information science and technology, (4) space science and technology, and (5) nuclear science and technology. Within these five programs fifteen multidisciplinary research centers will be organized, each one of which is required to include researchers from at least one Russian university and at least one non-Russian university. Much of the research in all of the research centers will be conducted at the participating researchers' home institutions, but SkolTech will be the convening place for joint work and collaborative activity. MIT will also help SkolTech establish a center for entrepreneurship and innovation. The main idea behind SkolTech is to help Russia become better at commercializing technology and inspiring entrepreneurial innovation. In June 2012, SkolTech accepted twenty-one outstanding Russian students from fourteen different universities into its master's program. These students attended a four-week Innovation Workshop at MIT in August 2012.[9]

In early 2012, groups of Russian scientists and engineers, many of them connected to Skolkovo, were invited to come to the United States to make pitches for their inventions before American venture capitalists and angel investors. One such event, which I attended, was held at the Cambridge Innovation Center, next to MIT. There the Russians made presentations on their projects to increase the efficiency of national electric grids, to reduce the volume of waste from sewage systems, to reduce losses of valuable

natural gas in drilling operations, and to produce refrigeration by a magne-
tocaloric device. After their presentations the Russians were invited to the
"Venture Café," where they mingled with dozens of other aspiring entrepreneurs and venture capitalists. The effort was designed both to help the
Russians find financial support for their innovations and also to change
their mentality, to get them to think and act like entrepreneurs. This effort
was not a philanthropic one; everyone was looking for new ways to make
money.

Some progress is obviously being made, but in the opinion of many of the
skeptical Western venture capitalists in the audience, the Russian presentations suffered from the same flaws: a failure to take economic calculations
adequately into consideration (many good ideas are economic failures),
an ignorance of how their ideas compared to ones elsewhere addressing
the same problems, and very little understanding of how innovations are
brought to market. The quality of the science in the Russian projects was
often high, but the social and economic context in which they would have
to operate was poorly understood. The problem of Russia's "lonely ideas"
had been lessened a bit, but it was still present.

Are RUSNANO and Skolkovo Solutions or the Latest Spasms?

Many countries have attempted to jump-start entrepreneurial technology
by creating high-tech programs and centers, and most have failed. A large
literature analyzing these efforts exists; one recent author, Josh Lerner of
the Harvard Business School, has called them the "Boulevard of Broken
Dreams" and has described how countries like Malaysia, France, Dubai,
and Norway have squandered hundreds of millions of dollars on unsuccessful efforts to duplicate California's Silicon Valley or Boston's Route 128
high-tech corridor.[10] Still, a few, such as Singapore, Israel, India, and China,
seem to be having some success. Where will Russia's efforts fall on that
spectrum?

Russia's efforts do not seem auspicious, for several reasons. Both RUS-
NANO and Skolkovo are top-down government-directed efforts, while
much of the activity in Silicon Valley and the Route 128 corridor came from
the bottom up (to be sure, with much government help, particularly from
military contracts). Both the Scientific-Technical Council of RUSNANO
and the Scientific Advisory Council of Skolkovo are made up of very senior

scientists and administrators. The record of such people in predicting the future course of technology is not good. A little history here may be helpful: just eight years before the bicyclemakers Orville and Wilbur Wright flew their airplane, the president of the Royal Society in Great Britain, Lord Kelvin, said, "Heavier-than-air flying machines are impossible." Thomas Watson, chairman of IBM, said in 1943, "I think there is a world market for maybe five computers." And Bill Gates of Microsoft said in 1981, "No one will need more than 637 kb of memory for a personal computer—640 k ought to be enough for anybody." The Russian government's high-tech projects depend too much on such senior people and do not leave enough room for the young rebels who are often responsible for technological breakthroughs.

When MIT first considered working with the Skolkovo Foundation to create a university, it proposed duplicating the basic organizational structure of MIT, in the sense of having undergraduate, graduate, postdoctoral, and faculty research and education combined. The Russians turned down this suggestion and insisted that the new university—the Skolkovo Institute of Science and Technology—should be only a graduate-level institution. No undergraduates would be educated there. Yet undergraduates are often the people who are the most entrepreneurial. A striking fact is that many of the biggest high-tech companies in the United States were founded by people who dropped out of school as college undergraduates in order to pursue and eventually commercialize their ideas. They include Bill Gates (Microsoft), Larry Ellison (Oracle), Michael Dell (Dell), Steve Jobs (Apple), Mark Zuckerberg (Facebook), Jawed Karim (YouTube), and Paul Allen (Microsoft). Sergey Brin and Larry Page (Google) were also dropouts, but as graduate students, not undergraduates. This correlation is too strong to be accidental. By the time students reach the advanced graduate level they have often lost the zest for doing something entirely new that propelled them earlier. Doctoral students have to cleave to the system, jumping over the hurdles put in place by their seniors in order to get their advanced degrees and often forced to follow research lines recommended by their professors. Younger students at the undergraduate level are sometimes more original.

Another reason for caution about Skolkovo is the motivations of the different partners in the effort, especially the foreign partners. Why are companies like Intel, Cisco, and Siemens willing to invest large amounts of money in Skolkovo? Their interest, properly and appropriately, is in promoting their own companies, and to do that they want access to talent and ideas all over the world (they have similar agreements with many countries). They will probably succeed in serving their own interests better than they will in helping Russia become a high-tech powerhouse. No doubt Skolkovo will produce some innovative ideas; where are those ideas likely to find their industrial applications, in Russia, with its very weak investor climate and its legal and political difficulties, or in the home bases of these international companies? As Daniel Treisman recently remarked in an analysis of Russia's attempts to innovate,

What matters most for growth is not where new ideas first appear but where they are developed. And this depends less on the brainpower of scientists or the extent of state research funding than on the quality of the business environment.[11]

The World Bank in 2012 ranked Russia at 120th place (out of 183 countries) on its "Ease of Doing Business Index" (Russia ranked 178th and 183rd, respectively, on "Dealing with Construction Permits" and "Getting Electricity").[12]

Without a dramatic improvement in Russia's business environment, efforts such as Skolkovo are going to help foreign partners more than Russia itself. A prominent Russian economist recently observed that Russia lacks "end users who benefit from hi-tech production." He called for the state to institute "business-friendly policies."[13]

A similar tendency extends to the cooperation of MIT and other Western universities with Skolkovo. Faculty members in leading research universities in the West are always looking for new young talent, people they can add to their research teams at home. What will happen when an MIT professor offers a young Russian researcher whom the professor has come to know through cooperation at Skolkovo a postdoctoral grant at MIT, or a position on an MIT research team? The possibility of brain drain and resulting resentment on the Russian side is always present.

RUSNANO and Skolkovo are better qualified to identify outstanding technical talent than they are to judge market conditions. Fitting new technologies to the market is more the key to success in commercial technologies than is technical inventiveness. At Skolkovo, the foreign firms and universities are promising to teach the Russians about management and market analysis, but when a promising innovation appears, who is likely to find the best market niche for it, the Russians or the foreigners? And who will benefit the most?

At the St. Petersburg Economic Forum in June 2012, Anatoly Chubais, head of RUSNANO, admitted in a strikingly frank fashion that RUSNANO was losing large amounts of money and making poor choices in its effort to support nanotechnology.[14] He gave four reasons for these failures: (1) the managers of RUSNANO could not keep up with the pace of development of nanotechnology, (2) the business model was "mistaken," (3) the market did not correspond to expectations, and (4) the risks involved in science and technology development had been inadequately appreciated. Critics of RUSNANO immediately pounced on Chubais's confessions and pointed out that despite failing to achieve their goals, the top administrators of RUSNANO were nonetheless enriching themselves; in 2011 the seven top RUSNANO officials received 492 million rubles (about $16 million) in compensation.[15]

The three issues on which Skolkovo is most vulnerable to failure are commercial strength, intellectual property rights, and brain drain.

The stated goal of the Russian government's spending so much money on Skolkovo is to elevate the Russian economy from one dependent largely on extractive industries to one boosted by knowledge, by high technology. Since, as this book illustrates, so much of the success of commercial technology depends on factors outside the laboratory (politics, social barriers, investment climate, corruption, etc.) a micro-technical center like Skolkovo, however talented its researchers and students, is likely to have limited commercial success in Russian society at large.

Disputes over intellectual property rights almost surely will be disruptive to the Skolkovo enterprise. Let us assume, for the purpose of discussion, that a joint research team of MIT and Russian investigators comes up with an idea that has genuine commercial potential. Will all the intellectual property rights go to the Russian side, or will they be shared by American and Russian researchers? The agreement between MIT and Skolkovo devotes quite a few pages to this problem, but none of those provisions has been tested. According to Aleksei Sitnikov, vice-president of Skolkovo Tech, the Russian side in the negotiations originally wanted "all the rights," but the MIT side forced an agreement in which assignment of intellectual property rights will depend on where the research was done and who were the major contributors.[16] If the financial stakes are high, such an indefinite agreement will provide a battle ground for contesting lawyers. Misunderstandings and grievances are likely.

Finally, brain drain will haunt the projects. The Russian students at Skolkovo Tech are scheduled to spend long periods of time in the United States at MIT. All classes at Skolkovo will be taught in English. As these bilingual Russian students progress to become talented researchers, there is a strong possibility that the best of them will receive job offers in the West, either at MIT or elsewhere. Again, misunderstandings and grievances are likely.

The greatest flaw of the Skolkovo and RUSNANO projects is that both are attempts to improve technology without basically changing the society in which technology must develop. This is the same defect that has plagued modernization efforts in Russia for three hundred years: Russia's leaders concentrate on developing new technologies, not on reforming society in such a way that advanced technologies become self-developing and self-sustaining. What we have learned from Russia's history, if we did not already know it, is that successful technical modernization depends on the characteristics of the society in which it is attempted much more than it does on individual technologies, however modern those technologies may be at the moment of introduction. Without thorough social reforms making Russia a more open, receptive, free, and stimulating society, individual technologies will have only very partial modernizing effects. They will work for a while, then become obsolescent. Russian society as it is presently organized is not likely to rejuvenate those technologies on its own. Once again, the Russian government will have to save the situation by direct action from above. Russia has not yet escaped from its age-old trap of fits and starts.

20 How Russia Could Break Out of Its Three-Centuries-Old Trap

You want the milk without the cow!
—Leading MIT administrator in Russia in 2010, objecting to Russian desires to have high technology without social reform

Can Russia, after several centuries of trying to modernize itself in a sustainable way, finally solve its problem? In principle, of course it can. Other countries have done it. Japan modernized a traditional society in less than a century. More recently, South Korea turned the trick in about forty years. Both Japan and South Korea are major players today in international high technology in a way that Russia is not.

The Soviet Union prohibited entrepreneurial capitalism and was committed to an alternative economic and political system. The disappearance of that regime provides Russia with the greatest opportunity in its history to benefit from its creative scientists and engineers. But will it do so? The first step toward that goal is recognizing the magnitude of the problem.

The problem is not a scientific or technological one, it is a societal one. I can perhaps best illustrate this by telling a little story. On one of my trips to Russia, in 2010, together with administrators and engineers from MIT, our group of Americans discussed the problem of Russian backwardness with Russian administrators and scientists. The Russians kept asking us about ways in which they could catch up with research at MIT in such fields as nanotechnology, biotechnology, and computer science. A high MIT administrator started talking about the "entrepreneurial culture" of MIT, the commitment of even undergraduates to successful innovation, the close relationship between university research and private investors, the economic and legal network that stimulated and protected start-up companies, and the political order, which allowed room for dissenting voices.

The Russians kept coming back to the technology itself. How could they develop "the next best thing" in high technology? Finally, the MIT administrator blurted out in exasperation, "You want the milk without the cow!"

This statement illustrates the immense problem that Russia faces. In order to become a leading participant in global high technology it must reform its entire society. Its problem of excelling in high technology may be solvable in principle, but in practice, finding a solution is dauntingly difficult.

How does one change the mentality of Russians about "business," shifting to a view that a businessperson making money from an innovation may be an admirable citizen, one of the major contributors to a country's prosperity? How does one create a political order in which successful entrepreneurs are not feared by government leaders as rivals for power and influence but promoted? How does one create a society in which freedom of expression, geographic mobility, and economic independence are valued and protected? How does one establish a legal system in which judges are independent from political authority, intellectual property rights are protected, and people accused of a crime have a chance of being acquitted? How does one create an economic and political order in which investors are not only numerous, but willing to take risks on developing novel ideas? How does one overcome rampant corruption, an environment in which extortionists quickly focus on any business that looks profitable? How does one reform a research and education system so that it brings teaching together with research as a single operation and creates at the highest level of achievement intellectually prominent scientists who care about application and economic development rather than priding themselves on working in an ivory tower?

Thus, the problem that Russia faces is truly monumental. However, improving the situation is not only possible, it is happening. Patent and intellectual property laws have been adopted and are being tested in practice. Schools of management have proliferated throughout the country, to the extent that "menedzhment" is almost a fad. The Russian government preaches the necessity of technological innovation and provides large sums of money for technology parks and technology start-up foundations. Russian business publications endlessly discuss the need for "modernization" and hash over different approaches, such as "catch-up modernization," "liberal modernization," and "forced modernization."[1] Furthermore, Russia already has hundreds of thousands of "entrepreneurs" working in retail stores, small businesses, banks, and trading outlets, both legal and illegal. So far not many of them have turned their interests toward commercializing high technology, but the potential for such a turn is present. In the last few years I have met undergraduates at several technical universities in St. Petersburg and Tomsk who spoke of their desire to become technology entrepreneurs (interestingly, these changing attitudes seem to be strongest at technology centers outside Moscow).

So—can Russia break out of its centuries-old trap and become a country where existing intellectual and artistic achievements are combined with new technological innovations that are successful commercially?

There is an easy answer to this question and a harder and more feasible one. The easy answer goes like this:

So Russia wants to become a high-technology superpower. Here is what it needs to do: it should become a normal Western nation. That means establishing true democracy, protecting human rights, creating a legal system that protects both intellectual property and entrepreneurs, reforming its higher education system so that it combines research and teaching and allows nongovernmental technology research centers, cracking down on corruption, and, finally, respecting and honoring businesspeople who make honest livings by promoting new technologies.

At the moment, the full achievement of this prescription is difficult to imagine. It would contradict Russia's traditions and run counter to the interests of many powerful people in Russia today.

The most promising development in Russian high technology today is not Skolkovo or RUSNANO, the government modernization projects. By themselves these two initiatives resemble the spasms of modernization that have been sponsored intermittently by the Russian government for centuries, only to be dragged down into obsolescence by the stultifying Russian social, political, and economic environment. No, the potentially most invigorating stimulus to Russian technology today is, counterintuitively, the demonstrations that have recently occurred on the streets of Moscow and other cities. These people are, at least in important part, professionals, the rising middle class, who represent something totally new in Russian history.[2] Only this new middle class has the power to transform Russia from a country of subjects to one of citizens; only it might create a country in which creativity and excellence are not swamped by corruption and repression. If Putin is sincere in wanting Russia to modernize, he should allow these authentic New Russians lead the country toward a sustainable knowledge economy instead of continuing its traditional unsustainable "innovation-on-command" economy.

In the absence of such a basic change in Russian politics, another path, one that could be called "gradual improvement," is feasible. It means trying to reform each of the individual elements named above in a way that gradually makes Russia more hospitable to technological innovation. The protection of intellectual property can improve, educational reforms can occur, the dominance of the central government can soften, changes in attitude toward businesspeople can commence, restrictions on mobility can continue to erode, closer links between Western companies and universities and Russian ones can be forged.

The result of such piecemeal reform will no doubt be disappointing to those who wish to see Russia quickly rise to the level of the great technological powers of today, but an age-old deadly pattern might be softened: that of "fits and starts," moments when Russia rouses itself and develops a leading technology, only to be followed by periods of retardation and obsolescence, a cycle that has occurred again and again, as illustrated in this book. That fateful pattern can become less pronounced even if Russia is not an international leader in high technology across the board. All it has to do is keep up, and all these little steps can help enormously in staying abreast with what is going on in the rest of the world. In this incrementalist scenario Russia can emerge not as a world leader in high technology but as a striving participant in it, leading perhaps in a few areas, following in many others, but following closely in all. For the foreseeable future, Russia is likely to remain in its trap because of its refusal to become a normal democratic country, but its isolation can lessen and the fits and starts can be less abrupt.

Acknowledgments

To acknowledge all the people who have helped me in the writing of this book would require reviewing my entire adult life, from graduate school forward, to study at Moscow University, to many, many trips to Russia. The task is impossible. However, a few people and institutions stand out. First of all is my wife, Patricia Albjerg Graham, who has accompanied me on many of these trips, who also speaks Russian, and who is an expert on education and research funding. She is a personal and professional inspiration. She is capable of both helping me and poking fun at me, two functions of enormous value. Our daughter, Meg, possesses the same skills; both mother and daughter have been joys to be with. I dedicate this book to them.

At Columbia University, where I began the study of Russia and stayed on as a faculty member for many years, two of my early professors were exceptional for their support of my delving into Russian science and technology, Henry L. Roberts and Alexander Dallin. My subsequent study at Moscow University was supported by the Inter-University Committee on Travel Grants, which lives on today as the International Research and Exchanges Board, an organization that has also many times given me assistance. The Guggenheim Foundation, the Woodrow Wilson Foundation, the Sloan Foundation, the Ford Foundation, the John D. and Catherine T. MacArthur Foundation, the Carnegie Corporation of New York, the National Science Foundation, the Civilian Research and Development Foundation, the Institute for Advanced Study, the American Philosophical Society, and the National Endowment for the Humanities have all supported me at one time or another. I have been very fortunate to live in a country where private and government foundations are generous in their assistance to scholars. The Civilian Research and Development Foundation, with the support of the MacArthur Foundation, has sent me to Russia several dozen times, visiting many Russian universities and research establishments. Marilyn Pifer at CRDF deserves particular praise for her steady support for Russian and American scientific contacts and mutual research. Particularly valuable colleagues in this work have been Victor Rabinowitch, Marjorie Senechal, Gerson Sher, and Harley Balzer. A French colleague, Jean-Michel Kantor, was an enormous help in exploring Russia's great mathematical tradition (see Loren Graham and Jean-Michel Kantor, *Naming Infinity: A True Story of Religious Mysticism and Mathematical Creativity,* Harvard University Press, 2009).

The universities where I have taught—Indiana University, Columbia University, the Massachusetts Institute of Technology, and Harvard University—are examples of the type of universities combining teaching and research that have helped the United States immensely in creating and applying knowledge and which are now models for many countries, including Russia. At MIT, people who have given me great assistance include Walter Rosenblith, Donald Blackmer, Merritt Roe Smith, Rosalind Williams, David Kaiser, Rafael Reif, and R. Gregory Morgan. At Harvard University Everett Mendelsohn helped me immensely, both institutionally and personally. Everett and his wife, Mary Anderson, are among my and my wife's closest personal friends. Peter Buck, both at MIT and at Harvard, has been a great conversational companion. At Harvard, I have been wonderfully assisted by the History of Science Department and the Davis Center for Russian and Eurasian Studies. At the Davis Center Tim Colton, Terry Martin, Lis Tarlow, and Alexandra Vacroux have created an atmosphere where scholarship flourishes. Tom Simons, my office-mate at the Davis Center, allowed me to benefit from his extensive experience in Russia and elsewhere.

In Russia, dozens of people have assisted me. The list of supportive librarians, archivists, and colleagues seems endless and stretches back more than a half century. Although I visited and lived in the Soviet Union many times, and was declared persona non grata by the Soviet government for a few years for digging too deeply into science and technology there, I have never encountered personal hostility in the Soviet Union or Russia; on the contrary, friendships flourished and scholarship blossomed. Particular Russian friends have been Vitalii Starchevoi, Sergei Kapitsa, Anton Struchkov, Nikolai Vorontsov (and his entire family), Mikhail Strikhanov, Daniel Alexandrov, Valeria Ivaniushina, Dmitrii Bayuk, Larisa Belozerova, Oleg Kharkhordin, and Irina Dezhina. Some of these people are described in my semi-memoir, *Moscow Stories*. Irina Dezhina and I have written a book together, *Science in the New Russia*, and a number of articles. She is a personal friend and an expert on science and technology policy in Russia.

One of the joys of my life is the American Philosophical Society, a remarkable remnant of the French Enlightenment whose assumption is that every educated person should know something about the significant developments in every field of knowledge. Conversations there with people like Mary Patterson McPherson, Alexander Bearn, Howard Gardner, Sara Lawrence-Lightfoot, Mary and Richard Dunn, Purnell Choppin, and Hanna Holborn Gray have helped me become a much better educated individual.

Ike Williams of Kneerim, Williams & Bloom skillfully served as my agent for this book, and his associates, Kathryn Beaumont, Katherine Flynn, and Hope Denekamp, are people dedicated to helping authors do their best. John Covell of the MIT Press was of immense importance in supporting this book. I would like to thank Marjorie Pannell for expert copyediting and Deborah Cantor-Adams for her excellent work in shepherding the publication to completion.

The photographs in this book were supplied by J. Mitchell Johnson of Abamedia, ably assisted by his archivist in Russia, Victor Belyakov.

Loren Graham
Cambridge, Massachusetts

Chronology

1479 (approximately)	Establishment of the Moscow Cannon Yard, which became one of the most technologically advanced foundries for cannons and church bells.
1632	Establishment of the Tula armory, which has had a continuous history to the present day. At moments it prospered, at other moments it lagged behind. Many Western artisans (e.g., Andrei Vinius, John Jones) worked at Tula.
Late seventeenth century	Establishment of the German Suburb (Nemetskaia Sloboda), where Western technology was introduced to Russia. Today one of Russia's leading engineering schools, the Baumann Institute, is located in this region.
1697–1698	Peter the Great traveled in Western Europe and studied shipbuilding and other technologies. He also brought Western artisans back to Russia and sent Russians to study technology in Western Europe.
1834	Miron Cherepanov built the first locomotive on the European continent that was not imported from England.
1835–1837	The first railway was built in Russia, between St. Petersburg and Tsarskoe Selo.
1872	Aleksandr Lodygin applied for an "invention privilege" for an electric incandescent lamp,

	several years before Thomas Edison began research on such a lamp.
1877–78	Pavel Yablochkov illuminated the streets of Paris and London for the first time with his electric arc lamps.
1888	Mikhail Dolivo-Dobrovolsky developed the three-phase electrical generator and three-phase electric motor.
1894	Aleksandr Popov built his first radio. In 1895 he achieved a range of 600 yards, in 1897 a range of 6 miles, and in 1898 a range of 30 miles.
1904	Aleksei Krylov built a machine for the solution of differential equations.
1913	Igor Sikorsky built and flew the world's first four-engine airplane.
1916	Mikhail Bonch-Bruevich developed flip-flop relays based on an electronic circuit with two cathode ray tubes.
1922	Oleg Losev built working solid-state radio receivers and transmitters.
1923	Oleg Losev built a light-emitting diode.
1924–25	Yuri V. Lomonosov developed the first operational mainline diesel locomotive in the world.
1927	Georgi Karpechenko first created a new species through polyploidy speciation during crossbreeding.
1920's	Russian biologists helped create the "evolutionary synthesis" and first developed the concept of a gene pool.
1935	Vladimir Shestakov proposed a theory of electric switches based on Boolean algebra, two years before Claude Shannon's famous MIT master's thesis on same subject.

1938	Soviet airplane designers claimed 62 world records, including the longest, highest, and fastest flights.
1940	Russian physicist Valentin Fabrikant proposed a laser and in 1951 applied for an "author's certificate" for it.
1947	Mikhail Kalashnikov designed the AK-47, the most popular firearm in history.
1948–51	Sergei Lebedev and associates built the first electronic computer in continental Europe.
1957	The Soviet Union launched the world's first artificial satellite.
1961	The Soviet Union launched the first human into space.
1964	Soviet physicists Aleksandr Prokhorov and Nikolai Basov received the Nobel Prize for developing the laser, along with the American physicist Charles Townes.
2000	Zhores I. Alferov shared the Nobel Prize in Physics for invention of the heterotransistor.
2007	RUSNANO was established to exploit nano-technology in Russia.
2010	The Skolkovo Foundation was established for the creation of a Russian version of Silicon Valley.

Notes

Introduction

1. Walter Isaacson, *Steven Jobs* (New York: Simon & Schuster, 2011), 321.

2. See, for example, T. Ravichandran, "Redefining Organizational Innovation: Towards Theoretical Advancements," *Journal of High Technology Management Research* 10, no. 2 (2000): 243–274; V. A. Thompson, "Bureaucracy and Innovation," *Administrative Science Quarterly* 5 (1965): 1–20; M. H. Meyer and F. G. Crane, *Entrepreneurship: An Innovator's Guide to Startups and Corporate Ventures* (Los Angeles: Sage, 2011), xvii; and H. Carpenter, "Definition of Innovation," *CloudAve*, June 29, 2010.

Chapter 1

1. "Zapiski Grafa M. D. Buturlina," *Russkii Arkhiv* 36 (1898), pt. II, 418.

2. Iosif Khristianovich Gamel, *Description of the Tula Weapon Factory in Regard to Historical and Technical Aspects*, edited by Edwin A. Battison (New Delhi: Amerind Publishing, 1988), 1.

3. Edward V. Williams, *The Bells of Russia: History and Technology* (Princeton, NJ: Princeton University Press, 1985), esp. 52.

4. F. N. Zagorskii, *Andrei Konstantinovich Nartov, 1693–1756* (Leningrad: Nauka, 1969); M. E. Gize, *Nartov v Peterburge* (Leningrad: Lenizdat, 1988).

5. *Istoriia Tul'skogo oruzheinogo zavoda, 1712–1972* (Moscow: Mysl', 1973).

6. Gamel, *Description of the Tula Weapon Factory*, 6–8.

7. Merritt Roe Smith, *Harpers Ferry Armory and the New Technology: The Challenge of Change* (Ithaca, NY: Cornell University Press), 325–326 and passim. I would also like to thank Smith for his helpful conversations and for showing me his unpublished manuscript, "The Military Roots of Mass Production, 1815–1913."

8. Edwin A. Battison, "Introduction to the English Edition," in Gamel, *Description of the Tula Weapon Factory*, xxiv.

9. Ibid., xxii.

10. Merritt Roe Smith, "Eli Whitney and the American System of Manufacturing," in *Technology in America: A History of Individuals and Ideas*, ed. Carroll Pursell, Jr. (Cambridge, MA: MIT Press, 1982), 45–61.

11. Ibid., 47.

12. Ibid., 48.

13. See Battison, "Introduction," xii, and James Carrington et al., "Examination of Hall's Machinery," manuscript, January 6, 1827, in *A Collection of Annual Reports*, Chief of Ordnance, vol. 1 (1812–44) (Washington, DC, 1878).

14. Battison, "Introduction," xii.

15. Nathan Rosenberg, ed., *The American System of Manufactures* (Edinburgh: Edinburgh University Press, 1969); Charles H. Fitch, "Report on the Manufactures of Interchangeable Mechanism," in *Tenth Census of the U.S.: Manufactures II* (Washington, DC: U.S. Census Bureau, 1883), 611–645.

16. John Sheldon Curtiss, *The Russian Army under Nicholas I, 1825–1855* (Durham, NC: Duke University Press, 1965), 127.

17. Battison, "Introduction," xxv.

18. Smith, *Harpers Ferry Armory*.

19. Ibid., 323.

20. Ibid., 330.

21. Gamel, *Description of the Tula Weapon Factory; Istoriia Tul'skogo oruzheinogo zavoda, 1712–1972*; V. N. Ashurkov, *Gorod masterov* (Tul'skoe knizhnoe izdatel'stvo, 1958); M. I. Rostovtsev, *Tula* (Tula: Tul'skoe knizhnoe izdatel'stvo, 1958); V. Mel'shiian, *Tula: Ekonomiko-geograficheskii ocherk* (Tula: Priokskoe knizhnoe izdatel'stvo, 1968); V. Berman, ed., *Masterpieces of Tula Gun-Makers* (Moscow: Planeta, 1981).

22. See Steven L. Hoch, *Serfdom and Social Control in Russia: Petrovskoe, a Village in Tambov* (Chicago: University of Chicago Press, 1986), 30 and passim.

23. Ibid., 189.

24. Berman, *Masterpieces of Tula Gun-Makers*, 11.

25. Gamel, *Istoriia Tul'skogo oruzheinogo zavoda, 1712–1972*, 32–35.

26. Ibid., 52.

27. Rosenberg, *American System of Manufactures*, 7.

28. Ibid., 16.

29. Jake Rudnitsky and Stephen Bierman, "Exxon Fracking Siberia to Help Putin Maintain Oil Clout," *Bloomberg Businessweek*, June 14, 2012. For references to Soviet work on fracking in the 1950s and 1960s, see Thane Gustafson, *Wheel of Fortune: The Battle for Oil and Power in Russia* (Cambridge: Harvard University Press, 2012), 545, note 21.

30. C. J. Chivers, *The Gun* (New York: Simon & Schuster, 2010).

31. *Rossiiskaia gazeta*, February 28, 2012, 3.

32. *Moscow Times*, November 30, 2012, 5.

33. M. T. Kalashnikov, *Ia s Vami shel odnoi dorogoi* [Memoirs] (Moscow: Dom "Vsia Rossiia," 1999).

Chapter 2

1. J. N. Westwood, *A History of Russian Railways* (London: George Allen & Unwin, 1964), 38.

2. Richard M. Haywood, *The Beginning of Railway Development in Russia and the Reign of Nicholas I, 1835–1842* (Durham, NC: Duke University Press, 1969), 242.

3. Quoted in Merritt Roe Smith, "Becoming Engineers," manuscript, August 31, 1987, 32.

4. V. S. Virginskii, *Cherepanovy* (Sverdlovsk: Sredne-Ural'skoe izdatel'stvo, 1987); Virginskii, *Efim Alekseevich Cherepanov, 1774–1842, Miron Efimovich Cherepanov, 1803–1849* (Moscow: Nauka, 1986); Virginskii, *Zhizn' i deiatel'nost' russkikh mekhanikov Cherepanovykh* (Moscow: Izdatel'stvo Akademii Nauk SSSR), 1966.

5. L. T. C. Roit, *George and Robert Stephenson: The Railway Revolution* (New York: Penguin, 1984); Hunter Davies, *A Biographical Study of the Father of the Railways, George Stephenson* (London: Quartet Books, 1977); Michael Robbins, *George and Robert Stephenson* (London: Oxford University Press, 1966).

6. M. I. Voronin, *P. P. Mel'nikov: Inzhener, uchenyi, gosudarstvennyi deiatel'* (St. Petersburg: Gumanistika, 2003), 195–222.

7. Ibid. 195–222.

8. Theodore H. Von Laue, *Sergei Witte and the Industrialization of Russia* (New York: Columbia University Press, 1963).

9. See the excellent biography of Lomonosov by Anthony Heywood, *Engineer of Revolutionary Russia: Iuri V. Lomonosov (1876–1952) and the Railways* (Farnham,Surrey; Burlington, VT: Ashgate, 2011).

10. Ibid.

11. Ibid., 208.

12. Ellen Barry, "Between Putin and Merkel, There's a Chill in the Air," *New York Times*, November 17, 2012, A6.

Chapter 3

1. George Westinghouse, "Opasnosti elektricheskogo osveshcheniia," *Elektrichestvo* 4 (1890): 68.

2. For the mood and spirit of Russian science in the late nineteenth century, see Alexander Vucinich, *Science in Russian Culture*, vol. 2 (Stanford, CA: Stanford University Press, 1963), and Elizabeth Hachten, "In Service to Science and Society: Scientists and the Public in Late-Nineteenth-Century Russia," *Osiris*, 2nd ser., 17 (2002): 171–209.

3. L. D. Bel'lkind, *Pavel Nikolaevich Iablochkov, 1847–1894* (Moscow: Izdatel'stvo Akademii Nauk SSSR, 1962), 190.

4. Robert Field and Paul Israel, *Edison's Electric Light: The Art of Invention* (Baltimore, MD: Johns Hopkins University Press, 2010), 91.

5. Liudmila Zhukova, *Lodygin* (Moscow: Molodaia gvardiia, 1989), 156.

6. L. D. Bel'kind, *Pavel Nikolaevich Iablochkov, 1847–1894* (Moscow: Izdatel'stvo Akademii Nauk SSSR, 1962).

7. *La lumière électrique*, no. 6 (1882): 378–379.

8. A description of Lopatin's revolutionary activities can be found in Woodford McClellan, *Revolutionary Exiles: The Russians in the First International and the Paris Commune* (London: Frank Cass, 1979), esp. 118–124. See also L. V. Davidov, *German Lopatin: Ego druz'ia i vragi* (Moscow: Sovetskaia Rossiia, 1984).

9. Moisei Radovskii, *Aleksandr Popov* (Moscow: Molodaia gvardiia, 2009).

10. Ibid., 9.

11. Gavin Weightman, *Signor Marconi's Magic Box: The Most Remarkable Invention of the 19th Century & the Amateur Inventor Whose Genius Sparked a Revolution* (Cambridge, MA: Da Capo Press / Perseus Books, 2003).

Chapter 4

1. Igor Sikorsky, *The Story of the Winged S* (New York: Dodd, Mead & Co., 1941); K. N. Finne, *Igor Sikorsky, The Russian Years* (Washington, DC: Smithsonian Institution

Press, 1987); Dorothy Cochrane, Von Hardesty, and Russell Lee, *The Aviation Careers of Igor Sikorsky* (Washington, DC: National Air and Space Museum / University of Washington Press, 1989).

2. Scott W. Palmer, *Dictatorship of the Air: Aviation Culture and the Fate of Modern Russia* (Cambridge: Cambridge University Press, 2006), 15, citing GARF f.102 DPOO 1909, d. 310, l. 19.

3. Sergey Sikorsky (Igor Sikorsky's son), in conversation with Loren Graham, Russian Research Center, Harvard University, April 1, 1987.

4. "Sovetskaia poliarnaia aviatsiia pokoril Ameriku," *Nezavisimaia gazeta*, March 25, 2005.

5. Kendall Bailes, *Technology and Society under Lenin and Stalin: Origins of the Soviet Technical Intelligentsia, 1917–1941* (Princeton, NJ: Princeton University Press, 1978), 386.

6. L. L. Kerber, *Stalin's Aviation Gulag: A Memoir of Andrei Tupolev and the Purge Era*, edited by Von Hardesty (Washington, DC: Smithsonian Institution Press, 1996).

7. Andrew E. Kramer, "At 35,000 Feet, a Russian Image Problem," *New York Times*, August 30, 2011, B6.

Chapter 5

1. Clifford G. Gaddy, *The Price of the Past: Russia's Struggle with the Legacy of a Militarized Economy* (Washington, DC: Brookings Institution Press, 1996); Fiona Hill and Clifford Gaddy, *The Siberian Curse: How Communist Planners Left Russia Out in the Cold* (Washington, DC: Brookings Institution Press, 2003).

2. Kendall Bailes, *Technology and Society under Lenin and Stalin: Origins of the Soviet Technical Intelligentsia, 1917–1941* (Princeton, NJ: Princeton University Press, 1978).

3. Loren R. Graham, *The Ghost of the Executed Engineer: Technology and the Fall of the Soviet Union* (Cambridge, MA: Harvard University Press, 1993). Material in the following section is taken from this book. See also I. A. Garaevskaia, *Petr Pal'chinskii: biografiia inzhenera na fone voin i revoliutsii* (Moscow: Rossiia molodaia, 1996).

4. See his "Rol' i zadachi inzhenerov v ekonomicheskom stroitel'stve Rossii," GARF (State Archive of the Russian Federation), f. 3348, op. 1, ed. khr 695.

5. Pal'chinskii, "Zamechaniia po povodu prichin maloi podgotovlennosti k samostoiatel'noi rabote, davaemoi spetsial'nymi vysshimi shkolami molodym inzheneram, i o sposobakh izmeneniia takogo polozheniia," GARF, f. 3348, op. 1, ed. khr. 1, l. 40ff.

6. GARF, f. 3348, op. 1, ed. khr. 751, l. 2.

7. Anne D. Rassweiler, *The Generation of Power: The History of Dneprostroi* (New York: Oxford University Press, 1988).

8. Ibid., 45–47.

9. Stephen Kotkin, *Magnetic Mountain: Stalinism as a Civilization* (Berkeley: University of California Press, 1995); John Scott, *Behind the Urals: An American Worker in Russia's City of Steel* (Bloomington: Indiana University Press, 1989).

10. Palchinsky, "Gornaia ekonomika," *Poverkhnost' i nedra* 1, no. 29 (1927): 9.

11. Cynthia Ann Ruder, *Making History for Stalin: The Story of the Belomor Canal* (Gainesville: University Press of Florida, 1998). For a Stalinist justification, see M. Gor'kii, L. Averbakh, and S. Finn, eds., *Belomorsko-Baltiiskii kanal imeni Stalina: istoriia stroitel'stva 1931–1934 gg.* (Moscow: OGIZ, 1934).

12. GARF, f. 3348, op. 1, ed. khr. 717.

13. See Bailes, *Technology and Society under Lenin and Stalin.*

14. Komarov, *The Destruction of Nature,* 57.

15. Paul R. Josephson, *Industrialized Nature: Brutal Force Technology and the Transformation of the Natural World* (Washington, DC: Island Press/Shearwater Books, 2002).

16. Hill and Gaddy, *The Siberian Curse.*

17. Ibid., 56.

18. Christopher J. Ward, *Brezhnev's Folly: The Building of BAM and Late Soviet Socialism* (Pittsburg, PA: University of Pittsburgh Press, 2009).

19. *The Great Baikal-Amur Railway* (Moscow: Progress Publishers, 1977), 8.

20. V. Perevedentsev, "Where Does the Road Lead?" *Current Digest of the Soviet Press* 40, no. 46 (1988), from *Sovetskaia kul'tura,* October 11, 1988, 3.

21. *The Great Baikal-Amur Railway,* 1.

22. *Current Digest of the Soviet Press* 39, no. 23 (1987) from *Pravda,* June 11, 1987.

23. *Current Digest of the Soviet Press* 41, no. 17 (1989), from *Pravda,* April 26, 1989, 3.

24. Ibid.

25. Ibid.

26. *Current Digest of the Soviet Press* 39, no. 10 (1987), from *Sotsialisticheskaia industriia,* February 11, 1987, 2.

27. V. Khatuntsev, "Why the Young Main Line Is Not Operating at Full Capacity," *Pravda,* June 11, 1987; *Current Digest of the Soviet Press* 39, no. 23 (1987), 21.

28. See Boris Komarov, *The Destruction of Nature in the Soviet Union* (White Plains, NY: M. E. Sharpe, 1980), 116–127; and Ward, *Brezhnev's Folly*, 12–41; see also Douglas R. Weiner, *Models of Nature: Ecology, Conservation, and Cultural Revolution in Soviet Russia* (Pittsburgh, PA: University of Pittsburgh Press, 2000), and his *A Little Corner of Freedom: Russian Nature Protection from Stalin to Gorbachev* (Berkeley: University of California Press, 1999); and Paul R. Josephson, *Resources under Regimes: Technology, Environment, and the State* (Cambridge, MA: Harvard University Press, 2004).

29. Conversations with Vladimir Sangi, president of the Peoples of the North, Moscow, December 1990 and October 1991.

30. Perevedentsev, "Where Does the Road Lead?," 3.

31. Tat'iana Gurova and Aleksandr Ivanter, "My nichego ne proizvodim," *Ekspert*, November 26–December 2, 2012, 19–26.

Chapter 6

1. A. G. Ostroumov and A. A. Rogachev, "O. V. Losev—pioner poluprovodnikovoi elektroniki," in *Fizika: problemy, istoriia, liudi*, ed. V. M. Tuchkevich (Leningrad: Nauka, 1986), 183.

2. Ibid., 183–217.

3. M. A. Novikov, "Oleg Vladimirovich Losev—pioner poluprovodnikovoi elektroniki (K stoletiiu so dnia rozhdeniia)," *Fizika tverdogo tela* 46, no. 1 (2004): 5–9 (available in English in *Physics of the Solid State* 46, no. 1 (2004): 1–4. See also "O. V. Losev—izobretatel' kristadina i svetodioda," http://led22.ru/ledstat/losev/losev.htm (accessed January 19, 2011).

4. "The Crystodyne Principle," *Radio News*, September 1924, 294–295.

5. O. V. Losev, "Deistvie kontaktnykh detektorov: Vliianie temperatury na generiruiushchii kontakt," *Telegrafiia i telefoniia bez provodov*, March 1923, 45–62.

6. Iu. R. Nosov, "Svet iz karbida kremniia," *Khimiia i zhizn'* 2 (2004): 42–46. See also Nikolay Zheludev, "The Life and Times of the LED: A 100 year history," *Nature Photonics* 1, no. 4 (2007): 189–192.

7. O. V. Lossev, "Luminous carborundum detector and detection effect and oscillations with crystals," *Philosophical Magazine* 5 (November 1928): 1024–1044; Losev, "Über die Anwendung der Quantentheorie zur Leuchtenerscheinungen am Karborundumdetektor," *Physikalische Zeitschrift* 30 (1929): 920–923; Losev, "Leuchten II des Karborundumdetektors, elektrische Leitfähigkeit der Krystalldetektoren," *Physikalische Zeitschrift* 32 (1931): 692–695; Losev, "Über den lichtelektrischen Effekt in besonderer aktiven Schicht der Karborundumkrystalle," *Physikalische Zeitschrift* 34 (1933): 397–403; Losev, *Telegrafiia i telefoniia bez provodov* 44 (1927): 485–494.

8. Zheludev, "The Life and Times of the LED," 191.

9. Egon E. Loebner, "Subhistories of the light-emitting diode," *IEEE Transactions on Electron Devices* 23 (1976): 675–699.

10. Novikov, "Oleg Vladimirovich Losev," 5.

11. *Telegrafiia i telefoniia bez provodov* 25 (July 1924): 342–343; *"Crystadyne" Home-made radio receiver using a crystal detector* (in English), circular no. 120 (Moscow: Bureau of Standards, 1925).

12. Loebner, "Subhistories of the light-emitting diode," 128.

13. Ibid., 685.

14. O. V. Losev, "Svechenie II: Elektroprovodnost' karborunda i unipoliarnaia pro-vodimost' detektorov," *Vestnik elektrotekhniki* 8 (1931): 247–255.

15. Ostroumov and Rogachev, "O. V. Losev," 212.

16. Kendall Bailes in his study of Soviet engineers in this period spoke of the "flight from production" for similar reasons. See his "The Politics of Technology: Stalin and Technocratic Thinking among Soviet Engineers," *American Historical Review* 79, no. 2 (1974): 445–469.

17. On the Ioffe Institute, see Paul R. Josephson, *Physics and Politics in Revolutionary Russia* (Berkeley: University of California Press, 1991); idem, *Lenin's Laureate: Zhores Alferov's Life in Communist Science* (Cambridge, MA: MIT Press, 2010); and his *Totali-tarian Science and Technology* (Amherst, NY: Humanity / Prometheus Books, 2005).

18. Lillian Hoddeson and Vicki Daitch, *True Genius: The Life and Science of John Bard-een* (Washington, DC: Joseph Henry Press, 2002), 276.

19. See the website http://www.tvr.by/eng/president.asp?id=69216 (accessed June 10, 2012). The Rice quotation is from Condoleezza Rice, "Russia's Future Linked to Democracy," CNN, April 20, 2005.

Chapter 7

1. H. J. Muller to O. Mohr, November 19, 1933, Lilly Library, Indiana University, Bloomington, quoted by Elof Carlson, *Genes, Radiation and Society: The Life and Work of H. J. Muller* (Ithaca, NY: Cornell University Press, 1981), 194.

2. Mark Adams, "The Founding of Population Genetics: Contributions of the Chet-verikov School, 1924–1934," *Journal of the History of Biology* 1, no. 1 (1968): 23–39; Adams, "Towards a Synthesis: Population Concepts in Russian Evolutionary Thought 1925–1935," *Journal of the History of Biology* 3, no. 1 1970): 107–129; Adams, "From Gene Fund to Gene Pool: On the Evolution of Evolutionary Language," *Stud-ies in the History of Biology* 3 (1979): 241–285; Adams, "Sergei Chetverikov, the

Kol'tsov Institute, and the Evolutionary Synthesis," in *The Evolutionary Synthesis: Perspectives on the Unification of Biology*, ed. Ernst Mayr and William Provine (Cambridge, MA: Harvard University Press, 1980), 242–278.

3. G. D. Karpechenko, "Polyploid hybrids of *Raphanus sativus* x *Brassica oleracea* L.," *Bulletin of Applied Botany* 17 (1927): 305–408. Karpechenko crossed a radish with a cabbage and produced a hybrid that gave fertile offspring.

4. S. S. Chetverikov, "O nekotorykh momentakh evoliutsionnogo protsessa s tochki zreniia sovremennoi genetiki," *Zhurnal eksperimental'noi biologii* 2 (1926): 3–54

5. N. L. Krementsov, *International Science between the World Wars: The Case of Genetics* (London: Routledge, 2005).

6. See David Joravsky, *The Lysenko Affair* (Cambridge, MA: Harvard University Press, 1979); Loren R. Graham, "Genetics," in *Science, Philosophy and Human Behavior in the Soviet Union* (New York: Columbia University Press, 1987), 102–156; and Valerii Soifer, *Lysenko and the Tragedy of Soviet Science*, (New Brunswick, NJ: Rutgers University Press, 1994).

7. Loren R. Graham, "The Biggest Fraud in Biology," in *Moscow Stories*, (Bloomington: Indiana University Press, 2006), 120–127.

8. See the website http://en.wikipedia.org/wiki/List_of_biotechnology_companies (accessed May 13, 2011).

Chapter 8

1. These and many other details are in Georg Trogemann, Alexander Y. Nitussov, and Wolfgang Ernst, eds., *Computing in Russia: The History of Computer Devices and Information Technology Revealed* (Braunschweig: Vieweg, 2001); see also L. G. Khomenko, *Dramatizm sudeb otechestvennoi kompiuternoi tekhniki i kibernetiki* (Kiev: Izdatel'skii dom Burago, 2003).

2. For the role an émigré American engineer played in developing Soviet microelectronics, see Mark Kuchment, "Active Technology Transfer and the Development of Soviet Microelectronics," in *Selling the Rope to Hang Capitalism?*, ed. Charles Perry and Robert Pfaltzgraff, Jr. (Washington, DC: Pergamon-Brassey, 1987), 60–77.

3. S. Frederick Starr, "New Communications Technology and Civil Society," and Seymour Goodman, "Information Technologies and the Citizen: Toward a 'Soviet-Style Information Society'?," in *Science and the Soviet Social Order*, ed. Loren R. Graham (Cambridge, MA: Harvard University Press, 1990), 19–50 and 51–67.

4. Anya Belkina, *System Preferences*, DVD, Boston, 2012 (Belkina is Rameyev's granddaughter).

5. For the atmosphere of Soviet physics in Stalin's time, see Alexei B. Kojevnikov, *Stalin's Great Science: The Times and Adventures of Soviet Physicists* (London: Imperial College Press, 2004); see also N. L. Krementsov, *Stalinist Science* (Princeton, NJ: Princeton University Press, 1997).

6. M. Iaroshevskii, "Kibernetika—'nauka' mrakobesov," *Literaturnaia gazeta,* April 5, 1952, 4.

7. Materialist, "Komu sluzhit kibernetika?," *Voprosy filosofii* 5 (1953): 210–219.

8. Slava Gerovitch, *From Newspeak to Cyberspeak: A History of Soviet Cybernetics* (Cambridge, MA: MIT Press, 2002).

9. A. I. Berg et al., eds., *Kibernetika na sluzhbu kommunizma* (Moscow, 1961).

10. Richard W. Judy and Robert W. Clough, *Soviet Computers in the 1980s* (Indianapolis: Hudson Institute, 1988).

11. Starr, "New Communications Technology and Civil Society" and Goodman, "Information Technologies and the Citizen: Toward a 'Soviet-Style Information Society'?," in *Science and the Soviet Social Order,* 19–50 and 51–67.

12. Loren R. Graham, *Moscow Stories* (Bloomington: Indiana University Press, 2006), 158–159.

Chapter 9

1. Theodore Maiman, *The Laser Odyssey* (Blaine, WA: Laser Press, 2000), 208.

2. Jeff Hecht, *Beam: The Race to Make the Laser* (Oxford: Oxford University Press, 2005).

3. I. G. Bebikh, ed., *Aleksandr Mikhailovich Prokhorov, 1916–2002* (Moscow: Nauka, 2004).

4. *Nikolai Gennadievich Basov* (Moscow: Nauka, 1982).

5. Maiman, *Laser Odyssey,* 7.

6. L. Biberman, B. A. Veklenko, V. L. Ginzburg, et al., "Pamiati Valentina Aleksandrovicha Fabrikanta," *Uspekhi fizicheskikh nauk* 161, no. 6 (1991): 215–218.

7. Maiman, *Laser Odyssey,* 208.

8. Quoted in Hecht, *Beam,* 14.

9. Nick Taylor, *Laser: The Inventor, the Nobel Laureate, and the Thirty-Year Patent War* (New York: Simon & Schuster, 2000).

10. Alexander Prokhorov, interview by Loren R. Graham, Moscow, fall 1986.

11. Joan Bromberg, *The Laser in America 1950–1970*, (Cambridge, MA: MIT Press, 1991); Hecht, *Beam;* and Taylor, *Laser.*

12. Maiman, *Laser Odyssey.*

13. I. A. Shcherbakov, "K istorii sozdaniia lazera," *Uspekhi fizicheskikh nauk* 181, no. I (January 2011): 71–78; A. M. Leontovich and Z. A. Chizhikova, "O sozdanii pervogo lazera na rubine v Moskve," *Uspekhi fizicheskikh nauk* 181, no 1 (January 2011): 82–91; I. M. Belousova, "Lazer v SSSR: pervye shagi," *Uspekhi fizicheskikh nauk* 181, no. 1 (January 2011): 79–81.

14. Maiman, *Laser Odyssey*, 186.

15. Ibid., 137 and elsewhere.

16. Iurii Medvedev, "Uspekh ili Nobel'," *Rossiiskaia gazeta*, December 13, 2007.

17. "Kogda teriaiut chest' i sovest'," signed by A. A. Dorodnitsyn, A. M. Prokhorov, G. K. Skriabin, and A. N. Tikhonov, *Izvestiia*, July 3, 1983.

18. Morris Pripstein, Physics Division, National Science Foundation, speech of February 15, 2010, American Physical Society.

19. Taylor, *Laser*, 287.

20. Laser Focus World, http://www.optoiq.com/index/photonics-technologies-applications/lfw-display/lfw-article-display.articles.laser-focus-world.volume-32.issue-7.departments.marketwatch.laser-industry-in-russia-struggles-to-build-market.html (accessed January 13, 2011).

21. *SPIE Professional,* July 2007, http://spie.org/x14793.xml (accessed January 17, 2011).

Chapter 10

1. See Keith Crane and Artur Usanov, "Role of High-Technology Industries," in *Russia After the Global Economic Crisis*, ed. Anders Aslund, Sergei Guriev, and Andrew C. Kuchins (Washington, DC: New Economic School, Peterson Institute for International Economics, Center for Strategic and International Studies, 2010), 95–123, 103. The Crane and Usanov article is excellent.

2. Asif A. Siddiqi, *Challenge to Apollo: The Soviet Union and the Space Race, 1945–1974* (Washington, DC: National Aeronautics and Space Administration, 2000).

3. Loren R. Graham, *Moscow Stories* (Bloomington: Indiana University Press, 2006), 18–21.

4. Robert MacGregor, "The Little Engine That Could," paper presented at Princeton University, spring 2011.The selection is part of his forthcoming Princeton University dissertation on rocket engines in the United States and USSR, 1945–1975.

5. Alissa de Carbonnel, "Botched Mars Mission Shows Russian Industry Troubles," Reuters, November 16, 2011.

6. Elena Shipilova, quoting Sergei Zhukov, head of space technology at the Skolkovo Innovation Center, in "What Role Will Russia Play in the Space Century?," *Russia Beyond the Headlines*, May 29, 2012.

Part II

1. For example, see Nathan Rosenberg, *Exploring the Black Box: Technology, Economics and History* (Cambridge: Cambridge University Press, 1994). For an analysis of failure more centered on business than technology, see Scott A. Sandage, *Born Losers: A History of Failure in America* (Cambridge, MA: Harvard University Press, 2005).

Chapter 11

1. *Istoriia tekhnicheskikh proryvov v rossiiskoi imperii v XVII–nachale XX vv.: Uroki dlia XXI v.?* (St. Petersburg: European University in St. Petersburg, 2010); Ingrid Oswald, Eckhard Dittrich, and Viktor Voronkov, eds., *Wandel alltäglicher Lebensführung in Russland: Besichtigungen des ersten Transformationsjahrzehts" in Skt. Peterburg* (Hamburg: LIT, 2002).

2. All above quotations are taken from *Istoriia tekhnicheskikh proryvov v rossiiskoi imperii v XVII–nachale XX vv.*

3. Deirdre McCloskey, *Bourgeois Dignity: Why Economics Can't Explain the Modern World* (Chicago: University of Chicago Press, 2010). On the rise of "prospecting" in eighteenth-century England, what we might today call "venture capitalism," see Robert C. Allen, *The British Industrial Revolution in Global Perspective* (Cambridge: Cambridge University Press, 2009).

4. The Old Believers were somewhat similar to Protestants, and they played an unusually prominent role in early Russian entrepreneurial activity. See William Blackwell, "The Old Believers and the Rise of Private Industrial Enterprise in Early Nineteenth-Century Moscow," *Slavic Review* 24, no. 3 (1965): 407–424.

5. See Alfred J. Rieber, *Merchants and Entrepreneurs in Imperial Russia* (Chapel Hill: University of North Carolina Press, 1982).

6. See Michael D. Gordin, *A Well-Ordered Thing: Dimitrii Mendeleev and the Shadow of the Periodic Table* (New York: Basic Books, 2004).

7. *Bol'shaia Rossiiskaia Entsiklopedia* (Moscow: BRE, 2006), 4:363.

8. See Loren R. Graham, *Science in Russia and the Soviet Union* (Cambridge: Cambridge University Press, 1992), 190–196, and idem, "How Willing Are Scientists to

Reform Their Own Institutions?," in *What Have We Learned About Science and Technology from the Russian Experience?* (Stanford, CA: Stanford University Press, 1998), 74–97.

9. Daron Acemoglu and James A. Robinson, *Why Nations Fail: The Origins of Power, Prosperity, and Poverty* (New York: Crown Publishers, 2012).

10. Yegor Gaidar, *Russia: A Long View*, trans. Antonina W. Bouis (Cambridge, MA: MIT Press, 2012), 153.

Chapter 12

1. See Kenneth Pomeranz, *The Great Divergence: China, Europe, and the Making of the Modern World Economy* (Princeton, NJ: Princeton University Press, 2001).

2. James E. Oberg, *Red Star in Orbit* (New York: Random House, 1981), 74–7; Leonid Vladimirov, *The Russian Space Bluff: The Inside Story of the Soviet Drive to the Moon* (New York: Dial Press, 1973), and A. A. Blagonravov, ed., *Uspekhi SSSR v issledovanii kosmicheskogo prostranstva* (Moscow: Nauka, 1968). See also Von Hardesty and Gene Eisman, *The Inside Story of the Soviet and American Space Race* (Washington, DC: National Geographic Society, 2007).

3. "New Broader Russian Treason Law Alarms Putin Critics," Reuters, November 14, 2012.

Chapter 14

1. Excellent recent surveys include Alain Pottage and Brad Sherman, *Figures of Invention: A History of Modern Patent Law* (Oxford: Oxford University Press, Oxford, 2000), and Christine MacLeod, *Inventing the Industrial Revolution: The English Patent System, 1660–1800* (Cambridge: Cambridge University Press, 2001).

2. *Manifest o privilegiiakh na raznye izobreteniia i otrkrytiia v khudozhestvakh i remeslakh*, June 17, 1812, PSZ 1830, vol. 32, no. 25143.

3. *Vysochaishe utverzhdennoe polozhenie o privilegiiakh*, November 22, 1833, PSZ 1834, vol. 8, no. 6588.

4. *Vysochaishe utverzhdennnoe mnenie gosudarstvennogo soveta ob izmenenii poriadka deloproizvodstva po vydache privilegii na novye otkrytiia i izobreteniia*, March 30, 1870, PSZ 1874, vol. 45, no. 48202.

5. A. Skorodinski, *Russian Patent Law and Practice* (London: Herbert Haddan & Co., 1911), 52ff.

6. A. Skorodinskii, *K peresmotru polozheniia o privilegiiakh 1896 goda na izobreteniia* (St. Petersburg, 1910), 110.

7. Anneli Aer, "Patents in Imperial Russia: A History of the Russian Institution of Invention Privileges under the Old Regime," *Annales Academiae Scientiarum Fennicae Dissertationes Humanarum Litterarum (Helsinki)* 76 (1995): 194.

8. Ibid, p. 69.

9. Ibid, p. 153.

10. Skorodinski, *Russian Patent Law and Practice.*

11. Aer, "Patents in Imperial Russia," 202.

12. V. I. Lenin, *Sochineniia*, 4th ed. (Moscow: Gosudarstvennoe izdatel'stvo politicheskoi literaturi, 1941), 22:263.

13. See N. A. Raigorodskii, *Izobretatel'skoe pravo SSSR* (Moscow: Iurizdat, 1949); B. S. Antimonov and E. A. Feishits, *Izobretatel'skoe pravo* (Moscow: Gosiurizdat, 1960); E. A. Maikapar, *Izobretenie i patent* (Moscow: Izdatel'stvo znanie, 1968); A. K. Iurchenko, *Problemy sovetskogo izobretatel'skogo prava* (Leningrad: Izdatel'stvo LGU, 1963); E. P. Torkanovskii, *Sovetskoe zakonodatel'stvo ob izobretatel'stve i ratsionalizatsii* (Kuibyshev: Kuibyshevskoe knizhnoe izdatel'stvo, 1964); V. A. Dozortsev, *Okhrana izobretenni v SSSR* (Moscow: Trudy TsNIIPI, 1967); V. P. Skripko, *Okhrana prav izobretatelei i ratsionalizatorov v SSSR* (Moscow: Nauka, 1972); "Polozhenie ob otkrytiiakh, izobreteniiakh, i ratsionalizatorskikh predlozeniiakh," *Voprosy izobretatel'stva* 10 (1973): 58–80.

14. James M. Swanson, *Scientific Discoveries and Soviet Law: A Sociohistorical Analysis* (Gainesville: University of Florida Press, 1984), 122.

15. Manfred Wilhelm Balz, *Invention and Innovation under Soviet Law: A Comparative Analysis* (Lexington, MA: Lexington Books, 1975), 125.

16. Joseph S. Berliner, *The Innovation Decision in Soviet Industry* (Cambridge, MA: MIT Press, 1976).

17. The Patent Law of the RF, coming into effect on Oct. 14, 1992, by resolution of the Supreme Soviet of RF of Sept. 23, 1992, no. 3517–1. Law of the RF "Concerning trademarks, service marks and the designation of places of origin of goods," of Sept. 23, 1992, no. 3520–1; Law of the RF "Concerning legal protection for computer programs and databases," of Sept. 23, 1992, no. 3523–1; Law of the RF "Concerning protection of the topology of integrated Microsystems," of Sept. 23, 1997; Law of the RF "On competition and restricting of monopolistic activity in goods markets"; Law of the RF "Concerning author's rights in contiguous rights" of July 9, 1993, no. 5351–1; Law of the RF "On selective achievements" of August 6,1993, no. 5605–1; Federal Law of the RF "On the ratification of the Eurasian patent convention," of June 1, 1995, no. 85-FZ; Civil Code of the RF, Part I, Federal law of Nov. 30, 1994, no. 5-FZ; Civil Code of the RF, Part II, Federal law of Jan. 26, 1996, no. 15-FZ; Tax code of the RF, Federal law of June 13, 1996, no. 63-FZ.

18. Edict (*ukaz*) of the President of the RF of May 14, 1998, no. 556, "On the legal protection of the results of R&D work of a military, special, and dual-use nature." Resolution (*postanovlenie*) of the government of the RF of Sept. 29, 1998, "On the first-priority measures for the legal protection of the interests of the government in economic and civil-law circulation of the results of R&D of military, special, and dual-use nature."

19. "Basic directions of government policy for the economic utilization of the results of S&T activity," Order (*Rasporiazhenie*) of the Government of the RF of Nov. 30, 2001, no. 1607-r.

20. See Derek Curtis Bok, *Universities in the Marketplace: The Commercialization of Higher Education* (Princeton, NJ: Princeton University Press, 2003).

21. See the entire June 2005 issue of *Technology Review*, published by the Massachusetts Institute of Technology, in which leading lawyers take opposite points of view on questions of open-source software and protection of digitalized information.

Chapter 15

1. Organisation for Economic Co-operation and Development, "The Knowledge-Based Economy," excerpt from *Science, Technology and Industry Outlook* (Paris: OECD, 1996), 3. http://www.oecd.org/science/sci-tech/1913021.pdf.

2. William Rosen, *The Most Powerful Idea in the World: A Story of Steam, Industry and Innovation* (New York: Random House, 2010).

3. Irina Dezhina, "Creating Linkages: Government Policy to Stimulate R&D through University-Industry Cooperation in Russia," manuscript, July 2012.

Chapter 16

1. Data from the website of Transparency International, http://cpi.transparency.org/cpi2011/results.

2. Russia Science Park Skolkovo Hit by Fraud Probe," Reuters, Feb. 13, 2013.

3. "Top Skolkovo Executives Threatened by Criminal Investigation," *East-West Digital News*, March 4, 2013.

Chapter 17

1. S. F. Ol'denburg, "Vpechatleniia o nauchnoi zhizni v Germanii, Frantsii i Anglii," *Nauchnyi rabotnik*, February 1928, 89.

2. The following section draws on Loren R. Graham and Irina Dezhina, *Science in the New Russia: Crisis, Aid, Reform* (Bloomington: Indiana University Press, 2008).

3. George W. Corner, *A History of the Rockefeller Institute: 1901–1953, Origins and Growth* (New York: Rockefeller Institute Press, 1964), 9.

4. F. Glum, "Zehn Jahre Kaiser-Wilhelm-Gesellschaften zur Förderung der Wissenschaften," *Naturwissenschaften* 18 (May 6, 1921): 293–300. See also Loren R. Graham, "The Formation of Soviet Research Institutes: A Combination of Revolutionary Innovation and International Borrowing," in *Russian and Slavic History*, ed. Don Karl Rowney and G. Edward Orchard (Columbus, OH: Slavica Publishers, 1977), 49–75.

5. S. F. Ol'denburg, "Vpechatleniia o nauchnoi zhizni v Germanii, Frantsii i Anglii" [Impressions of scientific life in Germany, France, and England], *Nauchnyi rabotnik*, February 1927, 89.

6. Denis Guthleben, *Histoire du CNRS de 1939 à nos jours* (Paris: Armand Colin, 2009); David Caute, *Communism and the French Intellectuals, 1914–1960* (New York: Macmillan, 1964), esp. 308–309.

7. Jonathan R. Cole, *The Great American University: Its Rise to Preeminence, Its Indispensable National Role, Why It Must be Protected* (New York: Public Affairs, 2009).

8. Corner, *A History of the Rockefeller Institute*, 39.

9. Corner, *A History of the Rockefeller Institute*, 541, and conversations with Alexander Bearn, former professor and trustee of the Rockefeller Institute and Rockefeller University, 2003–2005.

10. Conversations with Alexander Bearn, 2003–2005.

11. Richard Feynman, "The Dignified Professor," in *Surely You're Joking, Mr. Feynman! Adventures of a Curious Character* (New York: Bantam Books, 1986), 220–221 (iPad edition).

12. Roald Hoffmann, "Research Strategy: Teach," *American Scientist* 84 (January–February 1996): 20. See also Hoffmann's "University Research and Teaching: An Enriching and Inseparable Combination," *Boston Sunday Globe*, November 5, 1989.

13. David F. Feldon, James Peugh, Brianna E. Timmerman, Michelle A. Maher, Melissa Hurst, Denise Strickland, Joanna A. Gilmore, and Cindy Stiegelmeyer, "Graduate Students' Teaching Experiences Improve Their Methodological Research Skills," *Science* 333, no. 6045 (August 19, 2011): 1037–1039.

14. In 2010–2011 the Times Higher Education ranking did not include any Russian university in the top 200 in the world. In 2010 the Shanghai Jiao Tong University ranking listed Moscow University in 74th place. In 2013 Moscow University moved up to 50th place in the Times Higher Education World Reputation Rankings, but only after the rector of Moscow University, Viktor Sadovnichy, managed to get a

crucial question in the poll revised. The older question, which resulted in lower ranking for Moscow, was "Would you send a postdoc to Moscow State University?" The new question was "Would you be willing to accept a graduate of Moscow State University in your lab?" "Russia Beyond the Headlines," Advertising Supplement to the *New York Times*, March 20, 2013, p. 2.

15. Thomas Parkes Hughes, *Networks of Power: Electrification in Western Society, 1880–1930* (Baltimore, MD: Johns Hopkins University Press, 1983).

16. Theodore H. Maiman, *The Laser Odyssey* (Blaine, WA: Laser Press, 2000), 1, "Prologue."

Part III

1. Irina Dezhina and V. V. Kiseleva, *Gosudarstvo, nauka i biznes v Innovatsionnoi sisteme Rossii* (Moscow: Institut ekonomiki perekhodnogo perioda [IEPP], 2008); I. G. Dezhina and B. G. Saltykov, *Mekhanizmy stimulirovaniia kommertsializatsii razrabotok* (Moscow: IEPP, 2004); Harley D. Balzer, *Soviet Science on the Edge of Reform* (Boulder, CO: Westview Press, 1989).

Chapter 18

1. Alexey Eliseev, "Russian Investment & Massachusetts Technology: Winning Combination," speech before the U.S.-Russia Chamber of Commerce of New England, Boston, March 13, 2012.

2. Loren R. Graham and Irina Dezhina, *Science in the New Russia: Crisis, Aid, Reform* (Bloomington: Indiana University Press, 2008).

3. See Graham and Dezhina, *Science in the New Russia,* 116–125.

4. The program was administered by the Civilian Research and Development Foundation (CRDF).

5. "MacArthur Foundation Increases Commitment to Russian Higher Education," press release, November 11, 2009, MacArthur Foundation, Chicago.

Chapter 19

1. *Russia Beyond the Headlines*, advertising supplement to the *New York Times*, December 19, 2012, 4.

2. Despite the title, the following book is informative and helpful: David M. Berube, *Nano-Hype: The Truth behind the Nanotechnology Buzz* (Amherst, NY: Prometheus Books, 2006).

3. George Allen, "The Economic Promise of Nanotechnology," *Issues in Science and Technology,* Summer 2005.

4. See the website http://nanocolors.wordpress.com/2009//10?new-russian-nanotechnology (accessed spring 2010).

5. 139-FZ.

6. See the website http://crnano.typepad.com/com/crnblog/2007/05/russiaand-nano.html.

7. The above information came from Dmitry Akhanov, president and CEO, RUSNANO, USA, at a presentation at the U.S.-Russia Chamber of Commerce of New England, K & L Gates, LLP, Boston, Massachusetts, March 13, 2012.

8. Victor Vekselberg, "Onset of Russian-Chinese Collaboration in High Technologies," *China Daily,* June 5, 2012.

9. Press release, *Business Wire,* June 14, 2012.

10. Josh Lerner, *Boulevard of Broken Dreams: Why Public Efforts to Boost Entrepreneurship and Venture Capital Have Failed—and What to Do about It* (Princeton, NJ: Princeton University Press, 2009).

11. Daniel Treisman, "Russia's Tom Sawyer Strategy," *IWMpost* 106 (January–March 2011): 14.

12. See the website http://www.doingbusiness.org/rankings.

13. Andrei Bunich, "Russia Needs Hi-tech Demand at Home," *Russia Now,* July 9, 2012.

14. "Chubais chetvertoval Rosnano," MKRU, http://www.mk.ru/print/articles/720994-chubays-chetvertoval-rosnano.html (accessed July 5, 2012).

15. Ibid.

16. Alexei Sitnikov, presentation at the U.S.-Russia Chamber of Commerce of New England, K & L Gates, LLP, Boston, May 2, 2012.

Chapter 20

1. See the entire edition of the business magazine *Expert* (circulation 50,000) devoted to "modernization": *Ekspert,* no. 8, 2010.

2. Harley D. Balzer, *Russia's Missing Middle Class: The Professions in Russian History* (Armonk, NY: M. E. Sharpe, 1996).

Glossary of Names

Antonov, Oleg, 1906–1984 Soviet aircraft designer, and founder of the Antonov ASTC aircraft company. He designed many airplanes used domestically for agricultural and commercial purposes, as well as the enormous An-124 and An-225.

Alferov, Zhores I., 1930– Russian physicist, inventor of the heterotransistor, and corecipient of the 2000 Nobel Prize in Physics. He is currently cochairman of the Scientific Advisory Committee of Skolkovo.

Basov, Nikolai, 1922–2001 Russian physicist, recipient of Nobel Prize in Physics in 1964 for the development of the maser and laser.

Bonch-Bruevich, Mikhail, 1888–1940 Russian engineer. He developed electronic flip-flop relays in 1916 and did much research on radios and vacuum tubes.

Cherepanov, Miron, 1803–1849 Russian inventor who together with his father, Efim Cherepanov, built many steam engines, including a steam locomotive in 1834, the first steam locomotive on the European continent not made in England.

Chetverikov, Sergei, 1880–1959 Pioneer in developing the "modern evolutionary synthesis" combining Darwinian evolution with Mendelian genetics. Arrested in 1929, he was forced out of his position in 1948.

Chubais, Anatoly, 1955– Politician and business manager, advocate of privatization under Boris Yeltsin, and currently head of the Russian Nanotechnology Corporation (RUSNANO).

Dobzhansky, Theodosius, 1900–1975 Prominent geneticist, leader in developing the modern evolutionary synthesis. He emigrated to the United States in 1927 because of growing political difficulties in Russia and had a long career at the Rockefeller Institute (later Rockefeller University) in New York City.

Dolivo-Dobrovolskii, Mikhail, 1862–1919 Engineer and inventor. He developed a three-phase electrical generator and motor in 1888 and designed the first three-phase hydroelectric power plant in 1891.

Ershov, Andrei, 1931–1988   Soviet computer scientist and a pioneer in programming. He helped found the Novosibirsk Computer Center.

Fabrikant, Valentin, 1907–1991   First person to propose the concept of a laser, in his doctoral dissertation of 1939 and in an "author's certificate" of 1951.

Gagarin, Yuri, 1934–1968   Soviet cosmonaut, the first human being to go into outer space (1957). Gagarin became an international celebrity before dying in an airplane crash.

Gapontsev, Valentin, 1939–   Russian engineer and electronics entrepreneur. He left Russia in 1995, and emigrated to the United States, where he became head of the IPG Photonics Corporation.

Goltiakov, Petr, 1791–?   Russian arms craftsman who, together with his son, Ivan, made beautiful presentation arms at the Tula Armory in the nineteenth century for members of the royal family and other distinguished people.

Ioffe, Abram, 1880–1960   Prominent Soviet scientist, long head of the famed Leningrad Physico-Technical Institute, often called the cradle of Soviet physics. He did much work on semiconductors.

Jones, John   Master English gunsmith who went to Tula in 1817 and modernized the armory, using dies instead of the manual forging techniques previously used.

Kalashnikov, Mikhail, 1919–   Russian arms designer responsible for the original AK-47 assault rifle, the most famous small-arms weapon of the twentieth and early twentieth-first centuries.

Karpechenko, Georgi, 1899–1941   Biologist, creator of the first new species by means of polyploidy speciation during crossbreeding. He was executed by the Soviet secret police.

Kaspersky, Eugene, 1965–   Software designer and creator of Kaspersky Lab, a leading company in antivirus software.

Khodorkovsky, Mikhail, 1963–   Former Russian businessman, once the wealthiest man in Russia, and former head of Yukos, a large oil company. Arrested in 2003, he was imprisoned for a lengthy sentence.

Kol'tsov, Nikolai, 1872–1940   A creator of modern genetics. He was arrested in 1920, then released. He died on December 2, 1940, allegedly of a stroke, but possibly of poisoning. His wife committed suicide the same day.

Korolev, Sergei, 1907–1966   Lead Soviet rocket engineer (the anonymous "Chief Designer") and developer of Soviet spacecraft in 1950s and 1960s. He was imprisoned from 1938 to 1944 and held in a "sharashka," a special prison, where he continued space research.

Kurchatov, Igor, 1903–1960 Nuclear physicist, director of the Soviet atomic bomb project. He also played a large role in the construction in 1954 of the Obninsk nuclear power plant, the world's first such facility generating electricity for a power grid.

Krylov, Alexei, 1863–1945 Russian naval engineer who in 1904 built a machine for integrating differential equations.

Lebedev, Sergei, 1902–1974 Electrical engineer and computer scientist who in 1948–1951 built the first electronic computer in continental Europe.

Lodygin, Aleksandr, 1847–1923 Russian electrical engineer and inventor, one of the first developers of an incandescent lightbulb. He applied for an "invention privilege" for such a lightbulb in 1872, several years before Edison began work on lightbulbs.

Lomonosov, Iurii, 1876–1952 Railway engineer who in 1924 built what became the world's first operationally successful mainline diesel locomotive.

Lomonosov, Mikhail, 1711–1765 Russia's first significant scientist. He repeated Benjamin Franklin's kite experiment on electricity and was interested in practical arts such as mosaic making and ceramics, as well as in chemistry and physics.

Losev, Oleg, 1903–1942 Scientist and inventor. He built the world's first solid-state radios and anticipated the development of transistors and diodes.

Lysenko, Trofim, 1898–1976 Fraudulent biologist who, with Communist Party support, suppressed a whole generation of promising Soviet geneticists.

Mel'nikov, Pavel, 1804–1880 Talented railway engineer who constantly promoted, sometimes unsuccessfully, the early development of railways in Russia.

Mendeleev, Dmitrii, 1834–1907 World-famous developer of an early version of the periodic table of chemical elements and a promoter of technical industry, especially agricultural and cheese production and the petroleum industry.

Nartov, Andrei, 1683–1756 Russian inventor and craftsman, constructor of many lathes and mechanical devices.

Palchinsky, Peter, 1875–1929 Talented engineer who tried to help Soviet industrialization but criticized excessive top-down planning and ideological perversion of rational industrial planning. He was arrested for his criticisms and executed.

Peter the Great, 1672–1725 Ruler of Russia from 1682 until his death. A coercive modernizer of Russia, he brought many Western techniques to his country.

Popov, Aleksandr, 1859–1906 Russian physicist who was the first person to demonstrate the practical use of radio. He failed, however, as a commercial developer of radio.

Prokhorov, Aleksandr, 1916–2002 Russian physicist who in 1964, together with his student Nikolai Basov and the American physicist Charles Townes, received the Nobel Prize in Physics for the development of masers and lasers. In contrast to Townes, he did nothing to promote their commercial development.

Rameyev, Bashir, 1918–1994   A founder of Soviet computers following an independent architecture.

Serebrovskii, Aleksandr, 1892–1948   A pioneer in the development of Soviet genetics.

Shestakov, Victor, 1907–1987   Russian mathematician, logician, and theoretician of electrical engineering who proposed a theory of electric switches based on Boolean algebra two years earlier than the famed work of Claude Shannon in the United States.

Sikorsky, Igor, 1889–1972   Russian American pioneer of aviation who built in Russia the world's first four-engine passenger planes. Frustrated by politics, he moved to the United States, where he became most famous for his helicopters.

Timofeev-Resovsky, Nikolai, 1900–1981   Famed Soviet biologist who moved to Germany in the 1930s to escape Lysenkoism in the Soviet Union. Imprisoned by the Russians at the end of World War II, he worked in prison for many years, before eventually being released.

Tupolev, Andrei, 1888–1972   Pioneering aircraft designer. Arrested in 1937, he continued to design aircraft in a special prison camp ("sharashka") and was then freed during World War II. One of his best-known planes, the TU-104 (1955), was a very early jet passenger plane.

Vavilov, Nikolai, 1887–1943   Prominent biologist, known for his study of the origins of cultivated plants. Arrested in 1940 for his opposition to Lysenko, he died in prison in 1943.

Vekselberg, Victor, 1956–   Russian oligarch, controller of oil and aluminum industries. He is currently the leader of the Skolkovo innovation project.

Vinius, Andrei (Andrew)   Dutchman who established near Tula, south of Moscow, an armory that has existed continually to the present day.

Whistler, George Washington, 1800–1849   Prominent American railroad engineer who was employed in 1842 by Pavel Mel'nikov as a consultant for the building of the St. Petersburg–Moscow Railway. He died of cholera in St. Petersburg in 1849.

Yablochkov, Pavel, 1847–1894   Russian electrical engineer who first illuminated the streets of Paris and London by electricity.

Zworykin, Vladimir, 1888–1982   Pioneer of television technology.

Index